Development theory and the three worlds

Longman Development Studies
Edited by Professor D.J. Dwyer, University of Keele

Chris Barrow: Water resources and agricultural development in the tropics

Piers Blaikie: The political economy of soil erosion in developing countries

David Phillips: Health and health care in the Third World

Forthcoming

William Gould: People and education in the Third World

Frank O'Reilly: Energy and environment in the Third World

Development theory and the three worlds

Björn Hettne

Copublished in the United States with
John Wiley & Sons, Inc., New York

Longman Scientific & Technical,
Longman Group UK Ltd,
Longman House, Burnt Mill, Harlow,
Essex CM20 2JE, England
and Associated Companies throughout the world.

Copublished in the United States with
John Wiley & Sons, Inc., 605 Third Avenue, New York, NY 10158

First published 1990
Reprinted 1992

British Library Cataloguing in Publication Data
Hettne, Bjorn
Development theory and the three worlds. – (Longman development studies).
1. Economic development. Theories
I. Title
330.1

ISBN 0-582-00559-0

Library of Congress Cataloging-in-Publication Data
Hettne, Björn, 1939–
Development theory and the three worlds/Björn Hettne.
p. cm. – (Longman development studies)
Bibliography: p.
Includes index.
ISBN 0-470-21519-4 (Wiley)
1. Economic development. I. Title. II. Title: Development theory and the 3 worlds. III. Series.
HD75.H475 1990
338.9-dc20 89-12282
CIP

Set in Linotron 202 10/11 pt Plantin

Printed in Malaysia by PA

To Birgitta

Contents

Preface

This book is the result of a series of revisions starting with a SAREC (*Swedish Agency for Research Cooperation with Developing Countries*) workshop on development theory in 1977, where I presented a paper dealing with some key issues in development theory as the situation stood in the mid-1970s. In 1978 SAREC published a revised version in its report series, and again in 1982 an expanded version. Between these two revisions the title was bravely changed from 'Current Issues in Development Theory' to 'Development Theory and the Third World', indicating an ambition to give a more comprehensive overview of this new and fascinating field, and focusing particularly on attempts to overcome its Eurocentric bias.

The present version is even more ambitious, since it underlines the trend to see development theory as a set of theories relevant for understanding development problems in different geographical and historical contexts. Hence the title. There will surely be those who dispute the existence of three worlds, some arguing for a 'one world' approach, others pointing to the existence of even a 'fourth' and a 'fifth' world. Static concepts are necessarily imperfect reflections of a changing reality. I think that the 'three worlds approach' – with modifications – has relevance, although it may not survive the next revision. However, development studies as a specific social science tradition may in fact not survive for long. It is at least my belief that its function has been that of a catalyst, in forcing the excessively specialized and static social sciences to focus on development and change, history and future, to borrow from each other, and – as a consequence – contribute to a revival of the classical tradition of a unified historical social science. That prospect is what makes this field of academic specialization particularly rewarding for its practitioners.

SAREC provided financial support to this latest version as well, for which I am grateful. Karin v Schleebrügge was particularly helpful in getting this project through. I would also like to acknowledge valuable research assistance from Leif Ohlsson and helpful comments from participants in a seminar series on development theory at my home base, the Department of Peace and Development Research, Gothenburg University. Individual chapters have been tested at various international academic gatherings and I find it hard to express the real depth of my indebtedness to all constructive comments from colleagues.

A book is of course always written and rewritten in dialogue with others. This is, however, particularly true of this book which so much relies on the ideas of others. I have acknowledged my debt to friends and colleagues in earlier versions. Here I would like to thank Kari Polanyi-Levitt, Emanuel de Kadt, András Kráhl, Attila Agh, and John Toye for criticism and valuable suggestions. Denis Dwyer has been an encouraging editor and David Seddon made a thorough last review saving me from many, if not all, embarrassments. However, none of my critics probably thinks that the book is 'finished'. My view is that a book never is – but at some point you must let it go.

Björn Hettne
Gothenburg, November 1988

Introduction

The research territory

The process of societal change has always attracted theorists of various persuasions, even before the social sciences, as we now know them, were born. The emergence of modern social science took place as European society was in the midst of a process of transformation from 'tradition' to 'modernity'. This gave a distinct mark to the classics: the nineteenth century political economists and the founders of sociology and anthropology.

As the industrial system was consolidated in the twentieth century, the evolutionist perspective, common to all the classics, was replaced by functionalism and equilibrium theories, and the tradition of Grand Theory by an orientation towards specialization, ultimately leading to compartmentalization and positivism.

A renewed interest in the development of so-called 'backward areas' after the Second World War gave rise to theoretical and methodological concerns which bore a certain resemblance to classical social science. It is with this 'third phase' of social science development this book is mainly concerned. Development theory can in fact be seen as a corrective to the static and ethnocentric biases in mainstream social sciences. Its 'project' from this point of view is to reconstruct the classical approach on new foundations. What these foundations are is the content of the book. To the extent that this renaissance becomes fulfilled, development theory may be transcended. It is thus open to question whether development theory *as such* has any future. What is of great importance, however, is that social scientists are concerned with important social issues, such as 'development'.

Thus, development theory contains various social science approaches which have tried to tackle the problem of 'underdevelopment' and were significantly changed in the process. This applies to the interdisciplinary, normative and basically Third-World-oriented development theory that took shape in the postwar period. To the extent that theorizing about development was a more or less mechanical application of the main body of theory of whatever discipline to the countries in the Third World, under the assumption that the differences between the contemporary problems and the 'original transition' (Roxborough 1979: 1) was one of degree rather than of kind, it will be rather neglected here. It would fall outside what I conceive of as development theory proper, although even Eurocentric approaches played a catalytic role in the development of development theory as provocations. Therefore they belong to the family and cannot be completely disregarded. The same holds for the more recent wave of fundamentalist critique of the formative phase.

Development theory is a loose body of contending approaches which, in spite of contradicting each other, also presuppose each other. It should be emphasized that I do not conceive of development theory as a specific body of thought uniquely applicable to the Third World. Rather the variety of Third World situations represents a challenge for the social sciences in their treatment of change and transformation, making them – ultimately – universal and global.

Thus, development theory really took off only after the discovery that the problems of development in the Third World were specific and qualitatively different from the 'original transition'. It is my contention that this discovery (which was necessary for the rather trivial reason of excessive Eurocentrism) led to a gradual theoretical enrichment and that, to take a step further, development theory ultimately will prove to be of relevance also in the industrial countries, where automatic growth can no longer be taken for granted and 'development' now presents itself as a problem. Thus, in spite of the fact that development theory emerged from tentative attempts at understanding the problem of 'underdevelopment' from the point of view of the 'developed', it gradually acquired an increasingly universal quality, i.e. an authentic universalism in contradistinction to the false universalism that characterized the Eurocentric phase of development thinking.

In the course of its evolution development theory has consequently become increasingly complex and 'nondisciplinary'. It is therefore necessary to elaborate further on the problem of definition, although the reader may feel reluctant to digest another definition of a clearly overdefined phenomenon. My intention, however, is not to propose new definitions but merely to stress one important point: *There can be no fixed and final definition of development, only suggestions of what development should imply in particular contexts.*

Thus to a large extent development is contextually defined, and should be an open-ended concept, to be constantly redefined as our understanding of the process deepens, and as new problems to be solved by 'development' emerge. As Paul Streeten has observed, there is a solution to every problem, but also a problem to every solution (Streeten 1978: 34). Theorizing about development is therefore a never-ending task.

Under the concept of development theory I shall subsume

theories (there exists no single, generally accepted, development theory) of societal change, which attempt to integrate different social science approaches to the development problem. Development theory is more concerned with *change* than is typically the case with conventional social science disciplines, such as economics, sociology or political science, trapped as they still are in functionalism and comparative statics. Development involves structural transformation which implies cultural, political, social and economic changes. Development theory is therefore by definition *interdisciplinary*, drawing on, but also questioning, many theoretical and methodological assumptions in both Marxist and non-Marxist social science.

Furthermore, development theory has from the start been closely related to *development strategy*, i.e. changes of economic structures and social institutions, undertaken in order to find consistent and enduring solutions to problems facing decision-makers in a society. Thus development strategy implies an actor, usually the state. The closeness between theory and strategy is due to the fact that development problems were defined as *national* problems and that, consequently, development theorists – particularly the pioneers – addressed themselves to governments.

As development theory originally grew out of a concern with so-called underdeveloped countries on the implicit assumption that the conditions in these societies were unsatisfactory and ought to be changed, it is more explicitly normative than social sciences in general. In the case of normative theory, the distinction between theory and strategy easily gets blurred. In positive theory on the other hand it is possible to make a clearer distinction, and to pose the question what strategy implications various theories would have and what roles different actors can play.

Development theory primarily refers to the academic or 'scientific' pursuit of knowledge, while we shall use *development thinking* as the more inclusive concept, referring not only to academicians, but also to ideas and views of planners, administrators, politicians, etc. Practitioners in development often rationalize and legitimize their policies in terms of more or less obsolete theory, in which case one should speak of *development ideology* rather than development theory. The same terminology is appropriate if a specific theory (and the strategy informed by it) is imposed upon a society by strong external forces without much regard for applicability. Thus, in one context, a certain set of ideas can play the role of theory, in another, the role of ideology. Consequently, distinctions between 'theory', 'strategy' and 'ideology' are in real situations not easily drawn. Many would, for example, regard the normative theories to be discussed in this book as development ideology. I believe the distinctions as such will prove useful, even if their application in some cases could be questioned.

The delimitation of 'development studies' from the rest of the social sciences is of course problematic, since the subject area has been institutionalized in so many different ways in different academic environments. Development studies can, as I understand it, be conceived of as a problem-oriented, applied and interdisciplinary field, analysing social change in a world context, but with due consideration to the specificity of different societies in terms of history, ecology, culture, etc. This 'internationalist' approach I see as one of the greatest assets in the development research tradition which today, however, generates more criticism than praise.

The approach of this book is that of *intellectual history*. For reasons of space the ideas are not always related to real developments, as they should be. It admittedly contains another bias in creating the impression that later theoretical innovations have replaced earlier ones. This is simply not the case, although new approaches sometimes were stimulated by shortcomings of previous theorizing. Of more importance, however, is the changing context of development thinking. Theoretical changes in the social sciences are different from, for example, physics, on which Kuhn based his theory of scientific change. In the social sciences 'paradigms' tend to accumulate, rather than replace each other, one reason being that they may fulfil ideological purposes, even after their explanatory power (if there ever was one) has been lost. Thus progress in the social sciences is to a very large extent a matter of subjective views and perspectives. Consequently the very structuring of this book reflects the prejudices of its author.

The first chapter takes a look at the contemporary scene, which for a decade or so has been characterized by a crisis both in development theory and in the 'three worlds of development': industrial capitalism, 'real' socialism, and the underdeveloped areas, which in their turn face rather different development problems. One important aspect of these crises concerns the role of the state, and whether the state forms part of the problem, or of the solution (or both).

'Crisis' is a difficult concept and a difficult subject as well. However, the late 1970s and the first part of the 1980s were dominated by crisis discussions, as far as the social sciences are concerned. Although this discussion has been characterized by a high degree of confusion, I nevertheless felt that it should set the scene for the book.

One way of finding a way out of the present confusion is to look back upon and critically examine the earlier conceptions of development and their transformations. The development of development theory has not been a smooth and evolutionary process. Rather it has been characterized by theoretical contradictions and ideological polarizations, at least after the pioneering years (late 1940s and early 1950s) were over. During this Eurocentric period theories were rather mechanically tried on Third World countries, of which many were colonies and of course never consulted about what kind of development they needed or wanted. We can therefore speak of the birth of development theory, only to the extent that doubts eventually were raised about the relevance of conventional social science theory in the 'new nations'. This debate started in economics and gave birth to development economics. Thus, the second chapter gives an overview of this Eurocentric phase of development theory, as well as early attempts to go beyond 'the special case'. It attempts to show to what extent development thinking was rooted in the concrete experience of Western economic history, and how this mode of thinking, characterized by endogenism and evolutionism, ultimately resulted in what came to be known as the *modernization paradigm*. We shall deal with its classical phase as well as with its more recent revival.

The third chapter is devoted to the intellectual revolution that took place in the mid-1960s, when the above-mentioned Eurocentric view on development was challenged, primarily by Latin American social scientists, and a theory dealing specifically with problems of *underdevelopment* rather than the natural history of development was born. The *dependencia* approach formed part of a general structuralistic orientation in development theory which also had its pioneers among development economists in the First World. The basic outlook can be summarized as follows:

Within a given structure, local, regional or international, there were certain positions which regularly and more or less automatically accumulated material and nonmaterial resources, whereas other positions were deprived of these resources. Development for one unit could therefore lead to underdevelopment for another, depending on how the two units were structurally linked. The conventional idea of development as merely a repetition of the economic history of the industrialized countries was gradually abandoned. To put it briefly, endogenism was replaced by exogenism. Closely linked to this perspective was the idea that the intellectual understanding of what development was about had been distorted by academic colonialism and that self-reliance also implied an indigenization of development thinking.

The two processes of self-criticism in the West, on the one hand, and indigenization of development thinking in the Third World, on the other, form, as I understand it, the obvious point of departure for a more comprehensive concept of development, reflecting the diverse global experiences of development processes.

Thus a tentative attempt is made to account for the process of indigenization of the social sciences, particularly development theory, in Latin America, Asia and Africa. By indigenization in this context I mean changes in problem-orientation, theory and methodology, making the social sciences more relevant to the Third World. Needless to say, this is a subject for a book in itself and my account is necessarily very selective, concentrating on one major example: the Latin American dependency school.

The *dependentistas* were primarily concerned with the effects of imperialism in peripheral countries, but implicit in their analysis was the idea that development and underdevelopment must be understood in the context of world capitalism. This view was further strengthened as the crisis of the 1970s showed the degree to which an interdependent world economy had become a reality. Dependence seemed to be a universal phenomenon, not only something which affected a special category of countries. Trends towards a more global analysis subsequently emerged in several social science traditions. It was also reflected in a new interest in global modelling as well as in international diplomacy oriented towards global reforms (the NIEO and the Brandt commission). The fourth chapter summarizes these different theoretical conceptions of the world as one system marked by an increasing degree of interdependence. It also draws some conclusions on the *globalization* of development theory for the fate of *national* development strategies.

Another trend within development theory arising in the mid-1970s was the stronger interest in the *content of development*, very often formulated in terms of 'another' or 'alternative' development. This normative line of thought, apparently stronger in the rich countries than in the poor, was a reaction to certain inherent trends in conventional paths of development or 'overdevelopment', leading to ecological imbalance and psychological alienation. The fifth chapter discusses some of these normative or utopian contributions and tries to estimate their significance.

The sixth chapter raises the question whether the experience of grappling with the problems of underdevelopment during a period of three decades has made development theory more relevant also for the so-called developed world. Are the recent efforts to apply development theory to European development problems a phase in the development of a more universally valid development theory? Has the industrialized world, for a long time the model for the 'underdeveloped', reached the limits of the mainstream model? How can this model be transcended and what are the alternatives? These questions are dealt with separately for Western and Eastern Europe. In the Third World there is a growing feeling that imitative development must come to a close, but the transformation of the original itself poses quite different problems.

Rather than making a synthesis, the seventh chapter concludes the book with a summary of the reorientations which can be identified in a field which admittedly is less well-structured today compared with when I made my first survey in the mid-1970s. I have nevertheless maintained more or less the same framework as was used in the 1982 book for the reason that I am still committed to the project of *one* – albeit pluralistic – research territory, in spite of the obvious disintegration and polarization of theoretical positions. In a way it is precisely this debate which makes it possible to conceive of development studies as a single field. For a debate to take place, there must be communication of some sort, and only through communication is it possible to transcend blocked positions and move towards convergence.

Chapter 1

Crises in development theory and in the world

> Just as the progress of a disease shows a doctor the secret life of a body, so to the historian the progress of a great calamity yields valuable information about the nature of the society so stricken. MARC BLOCH

Although this book is an exploration of the intellectual history of development theory, our main concern is to illuminate the contemporary dilemmas of development. In this chapter[1] we shall thus explore the elusive phenomenon of crisis; the crisis in development theory, the theoretical status of the concept itself, crises in the real world (or worlds) and, finally, the institutional crisis of the state, which seems to be a common denominator between the 'crises' in various regions of the world. The reader may find the crisis terminology somewhat excessive after quite a substantial amount of crisis analyses in the past decade. For this very reason, however, I think it is appropriate to begin this survey of development thinking with a discussion on 'crisis'. Furthermore, I think that the phenomenon of crisis, if an acceptable definition is found, could be an entry to fresh insights into the development process, as well as to new approaches in social science.

The chapter consists of three sections. The first deals with the current critique of development theory, particularly in terms of relevance, and discusses different theoretical approaches to the 'crisis', under the assumption that the unsatisfactory state of development theory is related to its inability to explain the 'crisis'.

The second section addresses the question whether the much discussed 'global crisis' is a homogeneous phenomenon, reflecting a single world system dynamics, or if the crises in the 'three worlds' are expressions of more or less autonomous regional and national dynamics. Rather than giving a straightforward answer to this question, which I think is somewhat premature, an effort is made to identify the specificity of the development problem in major world-regions, again taking 'crisis' as the point of departure.

The third section focuses on a specific institution, the state, and the basic question here is whether it is part of the problem or part of the solution with respect to the current developmental crisis. This discussion too, is by way of introduction. Of course we will have to return to these issues in later chapters.

1.1 Crisis of theory and theories of crisis

The contemporary development crisis is a challenge for development theory in many respects, one of them being that we do not really have a theory of the crisis.[2] Differently put: 'Crisis theory is itself in crisis' (O'Connor 1987: 158). Our incapacity to correctly understand the phenomenon of crisis in the context of the development process is an indictment of the social sciences in general and 'development studies' in particular.

1.1.1 The erosion of confidence

During the 1980s development studies was challenged by a fundamentalist, monodisciplinary trend in the academic world, and a neoconservative trend in politics. Both trends reduce the 'development problem' in a highly simplistic way, thus neglecting the insights achieved in the field during three decades of empirical and theoretical explorations into previously unknown territories. To this should be added persistent suspicions in Third World academic communities about the relevance of Western development research, suspicions that can only be reinforced by the trends just mentioned. These are the external challenges the discipline, if we may call it so, is facing.

Development studies experiences *internal* challenges as well, because many established truths and conventional wisdoms have been questioned and abandoned in the course of its development.[3] This has given rise to a wave of self-criticism and a 'decline of optimism' (Preston 1985: 166). To quote Robert Chambers 'it is alarming how wrong we were, and how sure we were that we were right' (Chambers 1985). In another retrospective essay John Toye makes the observation that real development, as opposed to the expectations of leading development economists of the 1950s and 1960s, contains a strong element of surprise not accounted for in development theory (Toye 1987b). Similarly David Apter, writing from the point of view of political science, observes: 'discrepancies between policies and concrete results have been so great and experiences so often disastrous that another round of thinking is called for . . .' (Apter 1987: 14). Unpredictability is of course a fact of life, but at least this much should be acknowledged by any social science which wants to be taken seriously.[4] In general the actual development experience, as compared to expectations in the formative period of development theory, i.e. the 1950s and 1960s, has been dismal in most so-called developing areas.

Put differently, the pioneering years were thus also a time of optimism, as Paul Streeten observes:

> It is not easy to convey, in the present atmosphere of gloom, boredom and indifference, surrounding discussions of development problems, what an exciting time of fervour these early years were. (Streeten 1981: 61–2)

Many examples can be given of this atmosphere of gloom, most of them referring specifically to development economics. In an essay entitled 'The Rise and Fall of Development Economics' Albert Hirschman complains:

> Articles and books are still being produced. But as an observer and longtime participant I cannot help feeling that the old liveliness is no longer there, that new ideas are ever harder to come by and that the field is not adequately reproducing itself. (Hirschman 1981)

Undoubtedly a certain amount of cynicism is creeping into the field. Any rethinking, however, is in itself a healthy phenomenon, if established concepts and theories are found to be deceptive and irrelevant. The debate in development studies during the last decade has, to my mind at least, been fruitful from the point of view of breaking through deceptive concepts and theories. This is a precondition for a relevant analysis of the crisis in the 'extramural' world. Thus the difficulty of dealing with different manifestations of the crisis conceptually, and of understanding how they relate, is in itself a crisis; a crisis in theory, or perhaps even a paradigm crisis. In the words of Stefan Musto, 'we have lost the navigational aid which a clearly defined paradigm would have afforded us' (Musto 1985).

Crisis implies that a certain arrangement no longer works, and therefore also the opening up of new opportunities and the need for new solutions in various fields of social life. To speak of 'crisis' rather than 'problem' normally suggests that a nonconventional solution is called for, i.e. a change in the pattern and direction of development. This is part of the definition of crisis but may always be contested since a crisis in this sense can be recognized only in retrospect. Otherwise, the question of crisis versus noncrisis is largely ideological. Radicals in favour of structural change tend to see 'crises' where defenders of status quo see problems of adaptation. The multidimensionality of the crisis and the discontinuities in the development process force us to take a new look at development theory.

The relevance of a discipline is of course partly dependent on its quality (theoretical relevance), but also on its relations with the centres of power (policy relevance). In the latter sense development studies has been increasingly marginalized due to the intellectual and political changes in the real world referred to above, although it is questionable whether it ever had a very strong position in political terms. Anyhow, the marginalization continues and development studies may end up as a fighting church.

We should not, however, forget the existence of an emerging 'development fatigue' among the development researchers themselves. Theoretical struggles, of which we have seen a lot, are tiresome. The lack of political impact necessarily breeds apathy. To this comes the painful policy failures of many of those 'adopted' developing countries, which tried to implement some of the favourite ideas held by development theorists in the 1960s and 1970s. In contrast some of the initially less admired countries have achieved unexpected breakthroughs in industrialization and economic growth. There is therefore an awkward silence, occasionally disturbed by extremely orthodox diagnoses – as if the development theory tradition, typically stressing the possibility of purposeful intervention and a broad understanding of development, had not at all existed. Development theory may not always have come up with the correct solutions on the policy level, but on the other hand the problems identified and analysed have not disappeared either. Instead, new problems have been added.

To speak of a crisis in development theory may suggest that this is a single field with a more or less homogeneous theoretical approach. In the concluding chapter it will be shown that this is by no means to be taken for granted.

1.1.2 The concept of crisis

One theoretical issue, which so far has not been satisfactorily dealt with in development theory, is the function of 'crisis' in the development process, and the actual theoretical status of the concept as such. In Europe, for instance, the era of sustained and more or less automatic economic growth seems to be over, unemployment has stabilized on a fairly high level (between ten and fifteen per cent in many countries) even during the more recent recovery and there is a frantic search for alternative life styles, as if people began to feel they missed the road somewhere. Perhaps the essence of the crisis, therefore, is doubt about the future of the Western or European model of development. The question is frequently raised whether this model is sustainable in the longer perspective, and 'crisis' seems to be an adequate term to the extent that this question is answered in the negative.

Naturally the 'crisis' is defined and understood in different ways in different parts of Europe. The concept is more commonly in use among the left and typically refers to problems in capitalist development.

An official Polish Encyclopedia gives this definition:

> . . . a serious breakdown in the process of economic growth in capitalism. Crisis is a phenomenon solely connected with the capitalist economies and does not occur in other socio-economic systems . . .
>
> (Quoted from Garton Ash 1985: 1)

This is indeed a narrow definition, since it excludes from the term important aspects of present day social reality in Poland. On the other hand there are in the current European situation few things that are *not* considered to be in a state of crisis. A certain precaution is therefore called for.

A distinction can first be made between the vernacular use of the term 'crisis' and different attempts to give it a theoretical status. In the first sense, amply reflected in current book titles, the term can be substituted by 'problem' without losing any meaning (Johnston and Taylor 1986). In the second sense 'crisis' means a critical juncture in an irreversible process at which a radical change has become necessary. Crisis is thus a period of transformation or 'transition'.[5] Of course only hindsight may reveal whether what today is referred to as 'crisis' will be seen as a crisis in this more precise meaning. This definition also leaves open the problem of delimiting the system supposedly in crisis and, of course, reaching a consensus on the diagnosis. Unless this is achieved, the 'solving' of the crisis will occur without conscious human action to influence the outcome, i.e. the transition will be more painful than necessary.

The problem of delimitation can be seen as part of the problem of reaching consensus. Is the 'crisis' global? Or does it make more sense to identify specific crises in different sectors (economy, security, ecology) in individual nation states? If we by 'crisis' refer to systemic crisis the concept has no meaning if the system we are talking about does not have a dynamics of its own. The 'interdependence' of the world of today makes it hard to grant this autonomy to 'national' economies. Therefore most theoretical schools which attempt to give the 'crisis' a scientific treatment tend to see it as a world-phenomenon. Furthermore they are typically more or less part of a Marxist tradition which retains the classical social science interest in change and transformation.

Mainstream Marxism treats crises in the capitalist mode of production as basically economic phenomena recurring in a cyclical pattern. During a capitalist crisis there will be a structural adaptation of the productive system, paving the way for a new period of expansion, until the autonomous dynamics within capitalism is completely exhausted. This in turn constitutes *the* crisis, which paves the way for socialism. In this tradition the phenomenon of crisis is situated in the system of production, or more specifically in the unavoidable contradiction between productive forces and social relations (Mattick 1981).

A consequence of the present world economic crisis and the failures to account for it is the renewed interest in economic long waves as a possible explanatory approach. To describe these cyclical movements is of course not to explain why they occur. There are in fact not one but several approaches, focusing on different causal factors such as entrepreneurship (Schumpeter), capital (Mandel), labour (Freeman) and raw materials (Rostow). N. D. Kondratiev is usually referred to as the originator of the idea, but there exist earlier versions (Freeman 1984). Long waves are analytically distinct from business cycles, although in reality they do of course coincide one way or the other.

The current debate on the reality and essence of the crisis often deals with the interpretation of the cyclical movements, whether they are short run or long run processes, and whether they signify a qualitative change or not. Since long waves usually are associated with technological change. new innovations and institutional adjustment, a theory of long waves will have to be a theory of crisis; crisis as a transition. Ernest Mandel, who foresaw the present crisis at an early stage (Mandel 1964), leaves the question 'transition to what' open. He would prefer socialism but does not rule out a 'fourth age' of capitalism (Mandel 1984).

During the development of 'historical capitalism' the stagnant B phase of a cycle has paved the way for an A phase of expansion. Since the crisis implies new openings and opportunities for other social projects as well, the crisis is characterized by military, political and ideological struggle, involving very different options today: a socialist world system, a neomercantilist world system, a 'greening' of the world, etc. The worldwide dimensions of the present crisis necessitate world-order solutions.

Among world-system theorists the idea of recurring cycles is central to the explanation of global dynamics. In their book *Dynamics of Global Crisis* Samir Amin, Giovanni Arrighi, Andre Gunder Frank and Immanuel Wallerstein (1982) all agree that the 'crisis' is worldwide and integral, and must be analysed as such (op. cit: 10).

But even these theorists, who are all more or less associated with the world-system approach, have different views about the interpretation of the world-system dynamics. In a concluding discussion the authors spell out their differences. Without going into much detail it can be noted that Frank and Wallerstein are more positive about the existence of a single world-system dynamics. They believe, for example, that the so-called long waves form part of the functioning of the system, whereas Amin and Arrighi emphasize the specificity of each wave and leave the future outcome more open. The latter are thus more 'voluntaristic' than are Frank and Wallerstein, who perceive the behaviour of the world-system in a somewhat mechanistic and deterministic way. This rather important difference leads to different views on a number of issues, for example whether the decline of the US is similar to earlier declines of Dutch and British hegemony, and whether the Soviet economic system is an integral part of the world division of labour governed by the rules of the capitalist world economy, or if it is more or less outside the world-system. They also differ with regard to the potential of self-reliance, or delinking, as a national strategy of development. Thus a world-system position, typically expressed by Frank and Wallerstein, would put heavy stress on the capitalist world economy as one integral system covering the whole globe and being moved by a single dynamics, completely overpowering the individual states that make up the system.

Other approaches to the 'crisis' take their point of departure in the national manifestations but without neglecting the role of emerging transnational forces. The so-called 'regulation school' with French roots explains the 'crisis' as a failure in regulation which leads to an accumulation crisis. By 'regulation' is basically meant the more or less institutionalized pattern of wage relations, investments, management of currency and economically motivated state intervention in general. Any 'regime of accumulation' is characterized by a 'mode of regulation' and the crisis essentially means the search for a new mode of regulation. The current accumulation regime is called 'Fordism', and thus the crisis is a 'crisis of Fordism'. It is explained by a change in precisely those factors which were essential to the postwar economic boom in North America and Western Europe: mechanized mass production coupled with mass consumption, high labour productivity coupled with relatively high working-class wages, welfareism and other forms of state interventionism. Today this accumulation regime is breaking down due to slowdown in labour productivity, international competition, enforced restructuring, deindustrialization and permanent unemployment, and political antiwelfareism. It does not seem to be the case that the various responses to the 'crisis of Fordism' as yet constitute a coherent new mode of regulation.

The 'regulationists' analyse the crisis primarily through the study of national specificities rather than developing a theory of a 'global regime of accumulation', although such a theory by no means is excluded (Liepitz 1987). Close to such a concept, particularly as far as regulation is concerned, is the idea of a 'world order' based on the hegemony of one-nation state, most recently the USA after the Second World War. The hegemonic power is the closest the anarchic international system can come to a world government and serves an important regulatory function in an integrated world economy. Therefore a decline in hegemonic power implies a 'world-governance' crisis, i.e. nationalist and protectionist policies challenging the existing rules of the game (Bretton Woods). Both world-system theorists and regulationists emphasize the role of declining US hegemony in the current crisis. It is of course not the decline of US dominance as such but the regulatory function of this dominance or hegemony which constitutes the crisis. From the world perspective the crisis is precisely about how this necessary function shall be performed, i.e. the search for a functional equivalent to hegemony.

The Frankfurt school puts more emphasis on the sociocultural sphere and provides an important complement to the approaches dealt with so far, since 'crisis' obviously also involves important subjective dimensions. To Jürgen Habermas a crisis can be both a crisis of efficiency and rationality on the level of the system, a crisis of legitimacy on the sociocultural level, and a crisis of motivation on the individual level (Habermas 1976). A society experiencing crises on all three levels (let us call this 'pervasive' crisis) is in a severe process of disintegration which may ultimately result in a breakdown. Obviously something has to emerge from such a breakdown, but the shape of what is to come is probably more complicated than Marx's famous birth metaphor indicates.

The advanced capitalist countries, characterized by the enhanced role of the state, face problems with system rationality and efficiency, and search for ways out of the crisis that will enable them to maintain a sufficient level of legitimacy and motivation. The contradiction in advanced capitalism is the problem of balancing these various demands and the result will be a legitimation crisis rather than an economic crisis. In other parts of the world the crisis may have a different pattern and a different focus, but the Habermas scheme is nevertheless, because of its multidimensionality, useful as a general framework and will be applied as such in the examples which follow.

The problem to which systemic level the theoretical concept of crisis should be applied still remains, however, and for the time being we leave it at that. In what follows we shall look into the specificity of the three 'worlds' of development as well as one particular institution, namely the nation state.

1.2 Three worlds of crises

The current economic crisis started in the late sixties or early seventies depending upon what indicators we use. But do we have sufficient evidence to speak of one single world crisis? That would imply that the present world is operating in accordance with some common logic and that to understand the crisis we have to understand that particular logic, whatever it is.

At a recent conference Louis Emmerij exclaimed: 'There is no World crisis! There is a European, a Latin American, and an African crisis!'[6] I am inclined to agree to a certain extent, but for methodological rather than theoretical reasons. The financial mess in Latin America, the starvation in Africa and the tendencies towards deindustrialization of Europe are different kinds of phenomena. Even if they can be seen as varying manifestations of the crisis, they can hardly be derived from or satisfactorily explained by a single world system logic. 'Crisis of global accumulation' may sound impressive as a concept but as an explanation of concrete manifestations of crisis in specific social contexts it resides on a metaphysical level.

In order to find more tangible explanations, one could relate the various crisis phenomena to systemic features in the three 'worlds' of industrial capitalism, socialism and underdevelopment, each of which with at least some common underlying dynamics. This is of course a very rough classification. The least homogeneous is the so-called Third World. Therefore we shall go further and look for more or less distinct regional problems in Latin America, Asia and Africa.[7] The time for disaggregation in development studies has come. The 'single basic cause' approach to underdevelopment disappeared with the decline of the dependency school, as did the unilinear approach with the decline of the modernization paradigm.

1.2.1 North: the welfare state questioned

The illusion that Europe somehow had 'arrived', that it had reached the highest stage of development, was shattered by the crisis of the 1970s. It became clear that there is in fact no state of 'being developed', only continuous processes of change, and these pass as 'development' or 'nondevelopment' depending on one's point of view.

The European crisis defies definition, and yet it was, and more or less still is, evident in many fields: trade, industrial development, employment and social welfare, integration, domestic politics, environment, security and defence, cultural values and lifestyles. To what extent do these manifestations form part of a world crisis and to what extent do they reflect autonomous and largely unrelated problems in European development? The answer to this question is dependent on our conceptual tools, since the crisis as such cannot be distinguished from our perception of the crisis. If crisis, as we suggest, can be a point of entry to the study of social transformation we would nevertheless need points of entry to the crisis.

As an angle from which to approach the specificity of the crisis in Western Europe, the welfare state may be an appropriate choice. It was a significant achievement, actually born or at least conceived in the midst of the previous major economic crisis, but long prepared by ideological currents such as Christian humanism, liberalism (of the J. S. Mill brand), conservatism (of the Bismarck brand) and socialism (of various brands). The major contribution in recent times is of course that of Keynes, who proved that the welfare state also made sense in economic terms. Sweden, where social democracy for decades has been the only serious political option, appeared as the most successful welfare state. Actually, welfareism and other forms of state interventionism are integral parts of modern capitalism, whether we call this 'late capitalism', 'Fordism' or 'the welfare state'.

To a large extent, the West European crisis in the 1970s and early 1980s has been a crisis for the welfare state. Even if a new era of growth were to open up, it would be difficult to return to the old project. The stagnating sectors have created unemployment, which is not sufficiently compensated for by the expansion of more dynamic sectors. At the same time the fiscal crisis for the state reduces the means whereby the employment crisis could be tackled in the Keynesian way (Mishra 1984). Furthermore, a strong political movement opposing the welfare state has undermined its legitimacy (Levitas 1986, Loney 1986, King 1987).

In countries where the public sector, and the social service sector in particular, is in a process of being cut down, a wave of criminality, political violence, racism and neofascism has become the constant worry of increasingly feeble and confused governments. Apart from these challenges, frightfully familiar to those who know some history, the European state apparatus also faces new types of social movement which are concerned about peace and ecology in a longer perspective rather than election politics of short duration and with little substance. The consensus on the welfare state (in its heyday far from being an exclusively social democratic concept) is breaking up without any clear indication of what it is going to be replaced with.

This is of course exactly what a crisis is all about. Some crisis managers want to strengthen the state further, others want to leave all decisions to the market, and a third current looks for the solution in a revival of (local) community. Since the first line seems to be the weakest we can speak of a crisis not only for the welfare state but for the state as such. Both market (neoliberal) and community (neopopulist) approaches contradict the logic of statism. At present the first trend seems to be the stronger.

The case of Britain may serve as an illustration.[8] The problem of widespread deindustrialization in that country is only one example of the general phenomenon of the decline of the traditional industries in Europe and the structural incapacity of the rather inflexible social system (the welfare state) to adapt to a dramatically changed world economy, shifting its dynamic core from the Atlantic to the Pacific. Thus, the problems of Britain are a mixture of more recent structural problems and its own secular decline which started a hundred years back, making it 'the most observed and analysed decline in modern history' (Gamble 1985). For this reason the causes and cures proposed are numerous, one major category relating to its global position, the other to its industrial competitiveness. Probably the key to understanding lies in linking them. British welfareism was not linked to expansion but to decline, and the solutions for Labour became the problems for Tories. In the 1970s when the world economic crisis set in a deadlock had been reached. The neoliberal response, to attack welfareism (termed 'socialism') and reduce the social service sector, in combination with monetarist measures against inflation which increase unemployment, obviously tends to raise the level of domestic social unrest. Growth has to be paid for by socioeconomic dualism.

In Britain this has been manifested in so-called soccer violence as well as racist clashes. The dynamics involved are dangerous, as the conservative analysis clearly separates the issue of social violence from the issue of unemployment, directly affecting almost three million. This number may sound alarming, but as long as the rest of the population can, if not increase, at least maintain their standard of living, the 'society' (relieved from the ideology of equality and solidarity) will not react. The unemployed are in fact sacrificed in the interest of continued consumerism for the majority. Thus, the unemployment-creating economic policy will continue, and the police force will be provided with whatever is deemed necessary to control the occasional outbursts of social unrest. The 1987 election clearly demonstrated the class cleavage which also was given a geographical expression: a growing South dominating over a stagnant North.

1.2.2 East: the fate of the socialist project

What about socialism then? In the eighties this concept (whatever the meaning) has little attraction in the West, compared to earlier periods. There are, however, certain indications of a rejection of socialism in the Eastern countries as well. This is a different type of crisis; a crisis for the credibility of the socialist project.

We could follow Bahro and make a distinction between the original idea of socialism and the 'actually existing socialism', suggesting a perversion of the socialist project, but also leaving open the option of a return to the original.[9] This implies, of course, the assumption that the idea of socialism was realizable in the first place. As Alec Nove puts the question: 'Does it make sense to "blame" Stalin and his successors for not having achieved what cannot be achieved in the real world?' (Nove 1983: ix).

However, there is a utopian vision behind each social project and this navigational aid gives to the social project its general direction. The 'crisis' is a matter of the degree of diversion from this course. To speak of a 'social project' instead of, for instance, 'the historical mission of the proletariat' implies that a project, such as the one of building socialism, may fail. No historicism is involved. Compared to the Soviet Union the Eastern European regimes face graver legitimacy problems, due to their shorter history and to the fact that these regimes were imposed by force (Lewis 1983).

Poland may be taken as a particularly sad example of project failure. The standard of living is going down, the people have lost faith in the project, and the erosion of the legitimacy of the state is shown by the simple fact that the military has taken over the leading role of the party. Two questions emerge: How could this happen? How representative is the Polish case for the socialist project at large?

The Polish crisis has been long in the making. The first popular outburst in 1956 produced the nationalist regime of Gomulka, whose legitimacy rested largely on being a victim of Stalinism. In 1970 he instead became the victim of the contradictions between consumerist demands and the need for a rationalization of the productive system, contradictions which the regime was incapable of solving. Gomulka was replaced by Gierek, who led the country to bankruptcy by simultaneously increasing both investment and consumption, financing it by foreign loans, made available by Western private banks during the period of easy credit. This policy was based on the concept of an 'open socialist economy'. In 1976, after a few apparently successful years, the third crisis was a fact, partly due to incompetent implementation, partly due to the global economic crisis of which Poland's overheated economy became victim. The time had come to pay for the white elephants. The economic growth cum consumer boom strategy had failed. Stern action was now politically impossible. In 1979 the national income began to fall and the following year the strike movement exploded. Poland had experienced a revolution of rising expectations and price rises were as usual met with strikes and demonstrations. Political authority in reality came to be 'shared' between the church, the party and the newly formed independent labour union (Solidarity) but there was neither shared responsibility nor joint action. The following years were consequently a drift towards disaster (Nove 1983: 146). In December 1981 martial law was declared and the state took command over society. Reforms were however still, or even more, an objective necessity. In 1987 a referendum was held, combining a package of political reforms with harsh economic measures. The power of veto had been transferred from the street to the ballot box. 'A more civilized way to say no', as some Poles expressed it.

The Polish crisis undoubtedly has its unique features but to a large extent it is possible to speak of a general crisis for the socialist project: diseconomies of scale, the 'new' class phenomenon, abuse of office and corruption, lack of participation, and above all the utopian attachment to a nonviable growth model. The hierarchical structure of the socialist societies in conjunction with their partly imposed, partly conscious and politically motivated isolationism makes the problem of change of direction (i.e. consistent and successful crisis management) rather difficult. At the same time the need for change is becoming more and more obvious (Werblan 1988). It is therefore most unfortunate for serious reformers that the New Cold War coincided with the manifest need for internal changes in the socialist countries. This made it possible for defenders of status quo to accuse the reformists of undermining socialism in the interest of imperialism. A major economic reform in the Soviet Union, which now seems to be on its way, may break the ice and force the hard-liners to reconsider, because that could hardly be interpreted as imperialist manipulations. Political change thus is a precondition for economic change.

This problem is an illustration to the point I am repeatedly making in this book, that development must be analysed in a political context. The economic logic of a situation does not always coincide with the political logic. However, it has been a persistent bias in development theory to abstract from other political concerns than those directly related to development. This is also why the problem of surprise referred to above has been overlooked. The 'discovery of surprise' is connected with the rediscovery of politics. Political considerations normally get the highest priority in designing development strategies. For instance, the imperative of military security has consistently been neglected as a factor behind the mainstream pattern of development. The centrality of security concerns for the decision-making centres in both poor and rich, socialist and capitalist, countries is consequently a serious limitation as far as alternative strategies are concerned.

1.2.3 South: the economics of survival

If we turn to the Third World the manifestations of crisis are even more dramatic in the sense that different problems overlap, giving the various crises a quality of pervasiveness. Thus, most problems appear in most regions. For the purpose of illustration we shall nevertheless focus on some dimensions which in the present situation seem particularly characteristic for the respective regions, starting with Latin America and the staggering debt situation. In Africa the crucial question today, after three decades of 'development' for most people is how to fill their stomachs. In Asia, at least in South and Southeast Asia, the threat has to do with another basic human need: the problem of identity. These are problems of survival: political, physical and cultural. But the roots to them all are economic.

Latin America and the debt problem

One of the reasons why development theory is in crisis is a complete change of the development agenda from long-run development issues to more immediate concerns, for instance debt management. The debt problem is affecting all regions, but it has reached truly staggering proportions in some countries in Latin America, where it has been considered a time bomb. The fear has been that a series of unilateral actions to suspend payments would trigger a collapse of the international banking system, a risk that subsequently has been diminished through various preventive measures (Griffith Jones 1988). The situation, however, is grave enough. It grew out of unprecedented international lending by private Western banks in the latter part of the 1970s. By the early 1980s the three largest debtors to the international banks were Mexico, Brazil and Argentina. During the period 1975 to 1981 loans to Argentina had increased by 615 per cent (Ferrer 1985). This foreign credit did not result in any increase in the country's productive capacity, instead much of it had been wasted in the import of weapons, luxury goods, capital flight and speculation. Hence the crisis.

The individual national histories of the huge Latin American debt accumulation vary somewhat, but the result is more or less the same. According to the World Bank average per capita real incomes in much of Latin America are back to the levels of the late 1970s. Thus many of these countries have lost a decade or more of development while putting huge obstacles in the way of their future development.

What gives the debt problem its dramatic character is the way it relates to the intensification of social tensions, and how this in turn affects the feeble democratization process initiated in Argentina, Brazil and Uruguay and now also in Chile. Due to the debt burden the capacity of the Latin American countries to design alternative policies is reduced. In order to be able to pay their debts, they are forced to implement the stern policies imposed on them by the IMF. This institution, which during the sudden flow of commercial bank credit in the 1970s became rather marginalized, is now back in the saddle and its conditionalities more politically controversial than ever (Payer 1985; George 1988).

The dilemma is thus that 'sound' economic policy is inconsistent with consolidating the political legitimacy, particularly of countries in the process of transforming disintegrating dictatorships into some kind of pluralist democracies. Latin America is not unique in this regard. The new democracies in Southern Europe face basically the same problem, but whereas the latter can use EEC membership as a way of stabilization, the former have to fight it out on their own. The balancing act is to live up both to people's expectations, raised by old and new populist policies, and to IMF conditionalities – at the same time!

It is the totality of and inherent contradictions in this situation that constitutes the development problem to be addressed by development theorists, not the relationship between inflationary rate, balance of payment and budget deficits, which is a purely economic problem. A society cannot be reduced to its economic sector and the economic sector cannot be reduced to impersonal market forces. Such a situation would imply some kind of Utopia and, as with all utopias, its implementation would presuppose total power. It has been tried in Chile, and the Chileans had to pay a terrible price.

In 1973 Chilean economic policy changed drastically following a military coup against the socialist government of Salvador Allende. It was a change from the typical Latin American developmental model, based on import substituting industrialization and state-interventionism, to a restoration of free operation of the market and a complete opening of the economy to the rest of the world, supported by military dictatorship. This has been referred to as the model of 'repressive monetarism' (Fortin 1984).

Political authoritarianism and repression was in fact an essential component of the strategy. Direct violence was needed to bring down the popularly elected government – and here structural violence in the form of destabilization was helpful. Violence was also needed to keep down a restive population suffering from the social effects of monetarist policy. Finally, violence was needed to eliminate all possible noneconomic factors that might intervene and disturb the operation of the market forces.

If one looks carefully into the implications of monetarism, the policy actually presupposes a ban on politics, which – since man is a political animal – implies a dictatorship. Only in Chile the policy has been carried that far. It is interesting to note that Pinochet discouraged political mobilization not only against the regime, but also in support of the regime. The ideal was the pure *cientifico*, reminiscent of the positivism of Auguste Comte which used to be so popular in Latin America.

It is obvious that monetarism (both in its technical sense and in the broader sense it is usually known) fails as development theory, although it may work as a growth strategy. The result is an extreme social polarization and external dependence, which must be considered when the development record is assessed. Development can never be understood by an analysis of the economic sector in isolation, however competent the analysis.

Africa and the food problem

In relative terms the debt problem has assumed dangerous proportions in Africa as well, but here starvation has for many years been front-page news. Many African countries face famine and drought. It may be argued that the very survival of the continent is at stake (Timberlake 1985). This, if anything, is a failure which should make theorists rethink development. A few observations on food crisis and development theory can be made here.

First of all it is probably true to say that development theorizing, like development praxis, until the more recent reversal in international aid circles suffered from an urban bias (Lipton 1977). Growth and modernization theory was inadequate from the point of view of the food problem, since it dealt with societal development in the abstract, as a process inherent in all societies and with a similar logic, regardless of the nature and structural position of the societies in question. Dependency theory was at least more explicit in dealing with the political dimension of underdevelopment. Still, the recommendations regarding development strategy were not concrete and specific enough, in particular as far as the food problem was concerned. However, we learned from the *dependentistas* how dependency structures forced countries away from the natural state of food self-reliance towards excessive specialization which led to the divorce of agriculture from nutrition.

Furthermore, the conventional approach to national security tends to reinforce growth and modernization strategies with emphasis on industrialization and corresponding neglect of agricultural development. To the extent that this policy of favouring the modern sector (including the export sector) will increase national security by increasing military capacity, the very same policy will create an internal security problem through the consequent urban/rural polarization, the neglect of subsistence production, and the resultant phenomenon of food riots. The concept of national security, therefore, should be redefined to include, among other things, food security.[10]

Referring back to the Habermas framework, there are in the Third World few crisis manifestations which threaten the legitimacy of established states more than food shortages and starvation. With admirable lucidity Amartya Sen has clarified that starvations are typically not direct consequences of food shortages but rather what he calls 'entitlement failures' (Sen 1981). This means that when people lose their food self-sufficiency through the process of modernization, they have to exchange their own labour power for food. If they fail in this respect they have nothing to hope for but a state sufficiently preoccupied with its own legitimacy to allocate money for food subventions. This is what a sensible state usually does, unless it detects more serious dangers elsewhere, for instance internal rebellions and external invasions. Unfortunately these types of threat to state survival seem to be on the increase. Thus international assistance to Africa is for survival rather than development, one more example of the changing development agenda.

The process of militarization is a serious obstacle to the achievement of food self-reliance, not only because of an unproductive use of scarce resources, but also because the military typically reinforces an unacceptable pattern of development, from the point of view of the food problem. For all these reasons the food problem is basically a political problem (George 1985).

Now let us turn to a country example. Ghana started out as one of the celebrated countries in Africa, with an early date of independence, a reasonably sound financial situation, and a strong political leadership in Nkrumah. Three decades later the country was struggling for survival under military rule, actually the third military regime, after five coups, six attempted coups and thirteen military plots (McGowan and Johnson 1984). Ghana proved to be one of the most unstable countries in Africa.

What went wrong, when and why? This is more difficult to explain than describing the immense contrast between now and then, although the country now seems to have reached the end of the decline, which was most evident from 1976 to 1983, and started an uphill path to consolidation. In terms of the Habermas framework the production system had come to a halt. Ghana lost not only its position as a leading cocoa producer but also the capacity to feed its own population. The low level of legitimacy is clearly shown in the figures of military intervention given above. On the level of motivation the public morale reached such low levels that an indigenous word for it, 'kalabule', became a general concept, necessary for understanding daily realities in Ghana during the period mentioned.

The fate of Ghana serves more than anything else as an illustration to the above-mentioned optimistic outlook of development economics in the late 1950s and early 1960s, as well as its later transformation into gloom. The basic weakness of Nkrumah's strategy was the imitative approach, the belief that simply a reproduction of those economic and social structures that are associated with Western 'modernity' automatically would lead to a 'takeoff into self-sustained growth'. However, it was consistent with development thinking of the time (Killick 1978).

Thus Ghana's crisis was initiated by Nkrumah, but there was no dearth of politicians to finish the job. Some regimes (like Nkrumah's) had high ambitions and failed, others were more interested in the prize of political power. On balance it seems that civil and military regimes have been equally responsible for the failures, although the second phase of Acheampong's military regime (1974–8) probably was the low mark. Jerry Rawlings's first coup (4 June 1979) was described by Rawlings himself as a 'house-cleaning' operation, which included both market women (blamed for the high prices) and the top leadership of the army (accused of more large-scale forms of looting). Eight senior officers were executed, among them three former heads of state.

Rawlings's second coup (31 Dec 1981) started a deep transformation of the political system, although it is difficult to tell what kind of revolution Ghana is going through (Ray 1986). The concept of crisis (or even 'pervasive crisis') with reference to the period from 1976 to 1983 seems to be appropriate, since Ghana is in a transition, becoming a completely different state from what it used to be.

Asia and the ethnic problem

In Asia neither the debt nor the food problem is absent – Bangladesh has a permanent food scarcity and South Korea has a substantial foreign debt – but the imminent crisis phenomenon today seems to be the ethnic challenge to the state. For this reason we shall illustrate the ethnic crisis with examples from Asia. The crisis is acute in South Asia and the problems are reappearing also in Southeast Asia, whereas East Asia is more homogeneous.

In Habermas's terms this is an identity crisis constituting a threat to national integration and with obvious implications for the process of economic development. However the causal relationship between the ethnic revival and development is at present not very clear. Thus another neglected dimension in development theory, of which we are now reminded by reality, is what Rodolfo Stavenhagen has called 'ethnodevelopment' (Stavenhagen 1986). In fact this is a completely new approach in development theory, challenging a lot of previous theorizing and conventional wisdom.

In modernization theory ethnic identity belonged to the traditional obstacles to development which were supposed to disappear in the course of development. Dependency theory looked for the external factors, and had little to say about ethnic problems. For Marxists this problem has traditionally been discussed in terms of 'the national question' which typically has been outside and beyond Marxist theorizing, which is what made it a 'question'. Thus it is easy to agree with Stavenhagen that development theory has ignored the ethnic question and failed to integrate it meaningfully into any analytical framework. One obvious reason for this is that development theory and social science theorizing in general is concerned with 'states' and 'national economies' as basic units, which precludes a serious treatment of the ethnic factor. Consequently the sudden rise of this factor has taken the social sciences by surprise. The ugly manifestations of the ethnic factor in South and Southeast Asia are particularly disturbing as the former has been one of the more stable Third World regions in political terms, whereas the latter has been one of the more successful in economic terms. As an example we can take a closer look at Sri Lanka, which is situated in South Asia but whose current development strategy is inspired by Southeast Asia.

In conventional development theory it is perfectly all right to speak of Sri Lanka's development problem, but it is rare to see references to 'Tamil' or 'Sinhala' development. Many aid agencies think of Sri Lanka as a receiver, not the Sinhalese or the Tamils. Rather the issue of who the actual receivers are in terms of ethnicity has been systematically avoided. Today when Sri Lanka emerges from civil war this convenient blindness is no longer possible.

To what extent was the ethnopolitical conflict caused by the 1977 change of development strategy from a regulated to an open economy? Even if ethnic rioting is an old phenomenon in Sri Lanka (or Ceylon) there was a marked increase in tensions after 1977 (Gunasinghe 1984). There is a link, obviously, but the actual causality is not easily established without detailed research about how the two types of development strategies affect different classes and sections with varying economic position within different ethnic groups. The point I want to make here is rather the conspicuous lack of that kind of analysis in the field of development studies.

However, it is very likely that a transformation of a regulated, protected nondynamic economy, where political patronage plays an important role, into a fluent, unpredictive and achievement-oriented economic system easily disturbs a sensitive relationship between ethnic groups, which have become used to exploiting specific niches in the economic system and come to regard these as a kind of natural right. On the whole Tamils were better situated to take advantage of the new rules of the game, the Sinhalese reacted violently, and the government, in spite of being the architect behind the economic change, seems to have encouraged the anti-Tamil riots of July 1983. This was the beginning of the civil war – and the end of the economic miracle. Only in 1987 did the two parties come to an agreement which still is to be implemented. The economic costs of the conflict have been enormous.

1.3 The state: problem or solution?

With some risk of committing the sin of overgeneralization, my argument is that a common feature of the crises in the different 'worlds' with their different systemic features is the crisis of the nation state as the 'normal' form of political organization:

- In the 'First World' of industrial capitalism there is, on the level of efficiency and rationality, a crisis of capitalist management (Keynesianism), which has led to the rise of a fundamentalist ideology, according to which there should be no state management at all, i.e. the market should decide about investment, production and consumption. This threatens the welfare system which so far has been the main instrument for the creation of political legitimacy.
- In the 'Second World', where the project of building socialism is carried out, there is a legitimacy crisis, because the living standards are decreasing, or at least not increasing with sufficient speed relative to the consumerist expectations that the socialist project unwisely chose as its basis of legitimation.
- In the 'Third World' the state apparatus has been in the hands of westernized élites, and in most cases no legitimacy whatsoever was created while some sort of development still took place. In the present debt, food and ethnic crises it is hard to see how the prospects for nation-building could be any brighter in the future. In Latin America the state is in full retreat and to the extent that it acts with a strong hand it is dismantling the state-as-planner structure. In Africa many states have for all practical purposes ceased to exist and local (more or less 'traditional') networks are filling the vacuum. In many parts of Asia the states are disintegrating due to emerging ethnic and territorial identifications.

The various challenges to the state mentioned here affect different dimensions of the 'state syndrome': from the state as the creator of the nation (the nation-building project) to the state-as-planner. It is not one problem we are facing but an enormous complex of old and new problems, ideological biases and theoretical confusions. All this makes it necessary also for development theorists to look more closely at the state, try to see what it is, what has been its role in development, what substitutes there are, and what implications all this has for development theory.[11]

1.3.1 Modernization and nation-building

The crisis in development theory is to my mind not resulting from theorizing having reached a dead end, which the present fatigue shown by the profession seems to suggest, but rather from the failure to seriously answer the old question: Whose development? From the very start development theorists – and development economists in particular – were addressing governments on the assumption that national development was to be given the highest political priority, and that their advice would be heeded. Furthermore, the state was seen as a homogeneous unit, autonomous from other agents, in possession of political and economic power, in control over external economic relations, and with the necessary techno-administrative and managerial capacity to implement plans (Gurrieri 1987).

Five-year plans now have become outmoded, however, and the market for development economists, at least as producers of plans, has shrunk drastically. In fact, very few governments in the Third World are concerned with development as a priority area. They are busy keeping themselves in power, fighting neighbours or suppressing rebellions. The nation state – the thing that development was all about – is challenged in many parts of the world. Beyond the vanishing state we begin to see the contours of a real but still somewhat indistinct world of people, struggling for human rights, livelihood and peace – in short the politics of survival. There are no common interests in development. This must be recognized by development theorists or the field will face its final demise.

Thus a basic weakness in development theory has been the erroneous assumption that development is a national goal that can be distinguished from other political goals and be given an exceptional and unquestioned position among them. For some reason we have assumed that the poorer and more backward a country is, the more its government will emphasize the role of development policy in order to 'catch up'. If development is defined as a general improvement in the standard of living, such an assumption is evidently wrong, and in retrospect perhaps also rather naive. In fact this became clear already with the so-called 'crisis of planning' in the mid-1960s (Faber and Seers 1972).

'Development' has at best really meant a strengthening of the material base of the state, mainly through industrialization, adhering to a pattern that has been remarkably similar from one country to another. This mainstream model has been enforced by the security interests of the ruling élite. In the mainstream model there are consequently potential conflicts, primarily between competing states within the interstate system, and secondly between, on the one hand, state power, and on the other, subnational groups challenging the legitimacy of the state. Thus, the concept of the nation-building project is a key to understanding what mainstream development has been all about – as well as its present crisis.

A nation-building project is unique for the simple reason that it uses a specific territory and a specific population living within that territory as 'building-materials'. However, any nation-building project contains common basic elements:

- The exclusive political/military control over a certain territory.
- The defence of this territory against possible claims from outside.
- The creation of material welfare and political legitimacy within this territory.

In order to succeed with such a project there is a need for an economic surplus which is created both internally and externally through participation in the world economy. Thus a development strategy is also a strategy for nation-building. The two cannot be separated. In terms of nation-building functions we could divide the surplus into three funds: a security fund, an investment fund and a welfare fund. The allocation of existing resources between these various funds differs according to the phase of nation-building and the various challenges facing it during a particular phase. Consistent failures in internal and external resource-mobilization, and continuous neglect of any one of these funds will weaken the very project of nation-building (Hettne 1984). That is why I call it a 'project'. Ultimately, it might not succeed.

hat bout payment fund?

1.3.2 The question of legitimacy

Development policy forms an integral part of the nation-building project.[12] By the creation and use of the welfare fund it is possible to achieve an integrated and consolidated nation state, characterized above all by the degree of legitimacy. In many Third World countries the movement towards internal coherence seems to have been interrupted, and neither the investment nor the welfare funds are being maintained. Instead these countries become increasingly militarized, i.e. the shrinking surplus is spent on 'security'.

In the long run, however, it is unlikely that even a substantial welfare fund can guarantee the legitimacy of the state. It is precisely because of the high level of welfare, and the consciousness which comes from improved education and increasing possibilities for cultural development, coupled with a capacity to understand the price of 'development', that people tend to develop identifications that transcend the nation. The great concerns today are global concerns. Therefore the nation state, at least as the exclusive form of political macro-organization, must be understood in historically relative terms and not as the final solution to the Hobbesian problem.

Thus types of predominant identification and types of state are related in increasingly complex ways. To simplify we can distinguish between two qualitatively different legitimacy crises for the nation state (see Fig. 1.1):

- One where the two development funds (investment and welfare) are exhausted and the state degenerates into a police state which people then try to escape, thus provoking further repression.
- The other where the high level of education and welfare – the results of a successful nation-building project – are transformed into transnational identifications and concerns embraced by a reasonably large segment of the population.

Fig. 1.1 Nation-building and alternative identifications

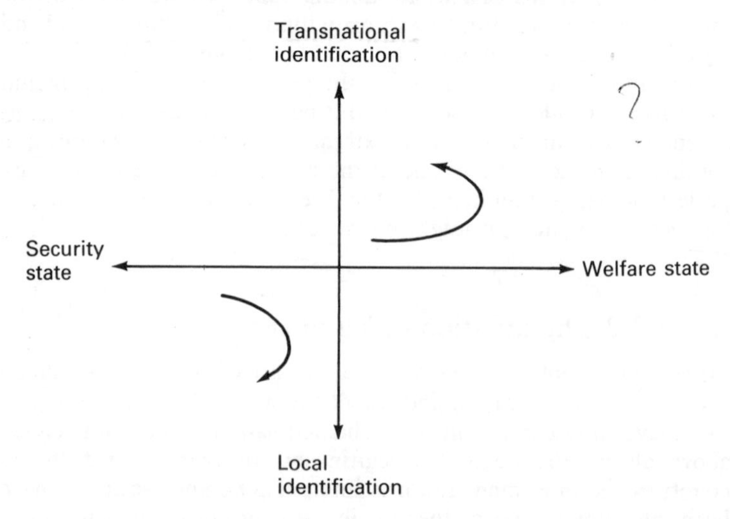

The latter types of identification are expressed for instance through the new social movements which due to their global concerns also challenge the legitimacy of the state. In the first case local and ethnic identifications will be strengthened; in the second global or transnational, but combinations of the two should not be excluded. In both cases the state will react by increased coercion, since the contradiction represents a fundamental threat to the nation state.

If the development of nation states really was a necessary path to progress, the African people in particular will have to pay a heavy price. According to South (Jan 1985) 25 states of the 51 members of the OAU are under military rule, 90 coups have taken place since 1960, the armed forces comprise 6.2 per 1,000 (to be compared with a Third World average of 3.4), the imports of arms 1976–80 amounted to 55.5 billion dollars (26.5 billion in the rest of the Third World) and more than 3.5 million people have been killed in uprisings, civil wars and massacres. In spite of all this suffering, the nation states do not seem to be on firmer ground today than when they emerged after decolonization.

As was pointed out above, many states in Asia as well are facing serious problems of disintegration. Pakistan split up once – and it could happen again. Lebanon for all practical purposes has ceased to exist as a state. Sri Lanka has been balancing between division and civil war. Are these traumatic events merely temporary obstacles in the way of the modernization project, or could it be that the era of demodernization has arrived? Is the state perhaps not the ultimate solution?

1.3.3 Beyond the nation state

Our discussion so far adds up to the question whether the nation state is an adequate political organization for taking the societies in the three 'worlds' out of the crisis and into a process of sustained development. This certainly can be disputed although the arguments would differ from case to case and no generalized conclusions can be made. Many states have undeniably brought 'their' societies into crisis.

Development in the form of industrialization has been patterned by the politico – military imperatives implied in the process of nation-building. Due to the imitative pattern of this process, industrialization results in a homogenization of world development creating international centre-periphery structures (where the structural positions of some countries may change over time). Due to the repetitive character of the industrialization process, there will be an excess capacity leading to periodical crises and a struggle for markets and resources (external conflicts). Due to the unavoidable social marginalization, enforced labour discipline, and misallocation of resources that accompany so many cases of Third World industrialization, there will also be repression and dictatorships (internal conflicts). The comparative advantage will be with the most aggressive and repressive nation states, at least until their own disintegration sets in. This fate should not be generalized but depends very much on the degree of original social cohesion within the respective countries.

As theoretical alternatives, responding both to the security problem and the development problem, we shall later discuss what usually is referred to as Another Development. Development strategies based on this approach would seem to be more 'peace intensive' than mainstream strategies. *Basic needs strategies* would reduce the need for internal repression, *self-reliance strategies* the need for international competition, *endogenous development* would create conditions for the cultural survival of aboriginal peoples, and *sustainable development* would eliminate tensions generated by resource scarcity. It could further be argued that Another Development would lead to a more secure society, not only in terms of less vulnerability, but also more creativity, confidence and consciousness, and a higher mobilizational potential (Hettne 1984). Rather than interdependence, an appropriate context for that kind of development would be a more regionalized (neomercantilist) world system consisting of more symmetrically related and self-reliant blocs (Hettne and Hveem 1988).

The main problem is of course how in practice a society characterized by 'peaceful' development will come about. In his book *Development and Underdevelopment* Gavin Kitching makes the following pertinent observation:

> An attractive utopian vision is not an adequate basis for a theory of development, nor does the desirability of a state of affairs guarantee its possibility.
> (Kitching 1982: 180)

On this point I would, however, like to point out that utopianism, defined as an emotional or opportunistic attachment to an unrealistic project, may characterize conventional as well as alternative models. The mainstream model is to my mind utopian because of its long term unviability, as indicated by the current crisis. This is more due to inherent contradictions of the model than to lack of political support for it. On the other hand the alternatives may be more viable in ecological or social terms, but here the accusation for utopianism rests on the lack of political backing. The non-viability of the mainstream model does of course not mean that all alternatives are equally realistic.

This is where the 'rediscovery of politics' comes in. It is a concept which must be qualified, since studies of the political dimension of underdevelopment are far from new. However, they have assumed a dominant role for the state in development to the neglect of other actors. As Rajni Kothari puts it:

> The assumption soon after the Second World War, by both elites and radicals, that the state would be a liberator and equalizer, is no longer avidly held and there is a creative reconsideration of the relationship between the state and civil society. There is a rediscovery of civil society as an autonomous expression of human and social will. (Kothari 1984)

Kothari holds that there is certainly still a need for theorizing but doubts whether it would be of much help, since 'the new actors on the scene will decide what their future will be'.

I, too, believe that a clear break with the view of development as a predetermined evolutionary process, imminent in the nation state or the world-system, is necessary and that the future should be conceived of as much more open. This necessitates an analysis of actors and their strategies or 'social projects' in a long historical world perspective (Friberg and Hettne 1985), or within particular regions, for instance Europe (Hettne 1986).

A useful framework for the categorization of various types of actors has been suggested by Marc Nerfin who associates the 'new politics', referred to by Kothari above, with what he calls the Third System.[13] Thus we can speak of the new political movements as 'Third System Politics', to be further discussed in Chapter 5.

Third System Politics is a type of political power not centred on the institutions of the state but a counter power negating the state. To the extent that the Third System is growing the role of the state is diminishing. The first question is whether this process actually takes place in the real world, and the second what the outcome might be. The first question was dealt with in the context of the nation state crisis, of which the growth of third system space is one manifestation. It is obvious that the state fulfils certain necessary societal functions and therefore has to be substituted as it no longer plays its proper role.

Alternative solutions are generally antistatist and usually have two, not necessarily contradictory, points to reference: the local community and the Earth. From those perspectives it is difficult to see how the state can be the crucial actor, except perhaps by delegating more power to other levels of decision-making: the global level, the (as yet only emerging) regional level and the (historically suppressed) local level. On the global and regional levels the states will obviously remain as important actors along with emerging transnational actors. The relevant actor on the local level would not be the state but issue-oriented social movements whose global aspirations transcend the nation state as the dominant mode of political organization.[14] The future is in fact more open than is commonly recognized.

Development theorists cannot be satisfied with only analysing ends and means in development as if there were a free choice between them. More attention should be given to the identification of development projects related to different groups and interests, and the analysis of the foundations of these projects in local, national and global power structures. From such an analysis we can, if not predict the future, at least draw certain conclusions about the future of development, about which development theory, regrettably, has very little to say. The crisis in development theory is clearly related to the crisis of the state, since development theory until recently has been based on the assumption of purposeful, positive intervention in the development process by the state. Reorientations in development theory flow from a reconsideration of this basic assumption.

Thus much work remains to be done. This must find a proper balance between global, holistic and interdisciplinary approaches on the one hand, and studies on economic sectors, individual countries and particular issues on the other. There is no other way in which the world dynamics can be grasped. Our inventory of the theoretical approaches to the crisis shows that the specificity of the analysis is higher, closer to the nation state level, at the same time as the analysis also reveals the lack of an autonomous dynamics on this particular level. In fact it may be argued that the crisis in development theory is a reflection of the disparity between the growing irrelevance of a 'nation state' approach and the prematurity of a 'world' approach.

A theme of this chapter has been that the present world crisis is a crisis for development theory, for the simple reason that the theory did not really prepare us for the crisis. Neither is it of much help in understanding the crisis, or the various manifestations of it in the three or more 'worlds'. Is development theory then useless? At least to this writer the development of development thinking in the last four decades is, as I hope to prove, a fascinating piece of intellectual history which, in spite of its shortcomings, will tell more about the concerns of these past decades to future generations, than histories of other social science specializations. In fact, I believe that development theory must be understood as a rediscovery of basic themes in classical social science related to change and transformation; that this process of rediscovery has only just started; and that the ultimate outcome may be a more unified social science. However, it will differ from its classical predecessor in two important respects: the evolutionist bias will have been replaced by a non-deterministic multilinear approach to development; and the

Eurocentric bias by a more authentic universalism, reflecting a global development experience.

There is still some work to be done before we reach that stage. Therefore the crisis in development theory (like, let us hope, in the real world) is not an end but a transition. Whether development theory will be called development theory, or simply social science, after that transition is less important.

Chapter 2

Eurocentrism and development thinking

The country that is more developed industrially only shows to the less developed, the image of its own future KARL MARX

We are fifty or a hundred years behind the advanced countries. We must make good this distance in ten years. Either we do it or they crush us.
JOSEF STALIN

By Eurocentric development thinking I mean development theories and models rooted in Western economic history and consequently structured by that unique, although historically important, experience. The first section of this chapter describes historical development patterns, seen as manifestations of a general Western development model. Then the rise of an economic development theory (development economics) in the 1950s and 1960s, reflecting these historical experiences but applied to the 'new nations', is traced. The third section summarizes different analyses of development within the interdisciplinary paradigm of modernization.

2.1 Development ideologies in western history

It is postulated that there exists a Western development paradigm which has coloured most Western contributions in this field, whether theoretical or practical. This may not appear as very persuasive for those who think of Western culture in terms of pluralism. One can, however, assume that there is a specific image of reality, underlying the multiplicity of views within Western culture (van Benthem van den Berg 1972). The task of making this picture more explicit is quite problematic due to the variety of perceptions about where Western society stands today and where it is heading tomorrow.[15] Let us first look back.

2.1.1 The European model in retrospect

The starting point for analysing the European development experience is the emergence of nation states and the constantly changing international context in which this took place. Here we must first consult the theorists of international relations. The dominant approach to international relations, i.e. to consider the international system as a form of 'anarchy', took shape during the 'modern' phase in European history, which roughly started with the peace of Westphalia (1648). This was the era of state-formation and nation-building, during which 'development' became a national interest, even an imperative.

To stress state-formation and nation-building as a historical process is also to say that it is difficult to tell at what particular date a state was born. The concept itself is not crystal clear. A state has been defined by Tilly as 'an organization employing specialized personnel, which controls a consolidated territory and is recognized as autonomous and integral by the agents of other states' (Tilly 1975: 70). These criteria should according to the same author be considered in a relativistic perspective, so as to make it possible to speak of a process of gradual increase in 'stateness', with reference to a particular political unit (ibid: 34). In Europe, the sixteenth century was a time of rising stateness, culminating in the later seventeenth century, the era of Absolutism. In the Third World, particularly in Africa, this is a more recent process, and its preconditions radically different. It is therefore advisable not to read the history of Europe into the future of the Third World. Thus, the relativistic view of stateness also implies the possibility of decreasing stateness, i.e. the disintegration of a state and the exhaustion of nation-building funds.

The rather drawn-out state-building process in Europe was a violent one and therefore people gradually learned to conceive their 'own' state as a protector and the rest of the world as an 'anarchy' and a threat to their security. Since then state-building has become a global process, and the state a universal political phenomenon.

One important reason behind the basic similarity between various national development processes is the military needs-related character of industrialization strategies. Even in the case of the 'first industrial nation' (Mathias 1969) the pattern of industrialization was significantly influenced by the 150 years of warfare that preceded the 'takeoff'. In the industrialization of the 'latecomers' the military needs were of course even more pronounced, as was clearly shown in the important role played by the state in the early process of industrialization. This process was consequently to a large degree imitative, and the fundamental pattern thus repeated from one country to another. Contrary to conventional modernization theory and orthodox Marxism, similarities in the pattern of economic growth did not reveal any inherent tendencies towards 'modernity' but political imperatives that made industrialization necessary for security reasons. To this should be added that the expansion of Europe was a competitive process involving a number of core-states struggling for hegemony.

In the era of the Great Wars, beginning in 1689 and ending in 1815, Britain was the dominant participant in international economic and political rivalries and moved towards ultimate hegemonic status. The Absolutist states that emerged in Europe during this period were 'machines built overwhelmingly for the battlefield' (Anderson 1975: 32). Peace was an exception in the era of Absolutism, and this legacy obviously also influenced the pattern of industrialization. Military demand for standardized output caused by the constant warfare and mass armies of the seventeenth, eighteenth, and early nineteenth centuries hastened factory production (Sen 1984). The difference between supplying large standing armies (with food, equipment, clothing, etc.) and the maintenance of decentralized feudal troops was of course enormous.

The military impact, however, related not only to demand factors but also to organizational and technological change. According to Sen the advancement of organizational methods occurred on two levels: first, the mass character of military demand stimulated the rationalization of the production process; secondly, the army itself provided a model for industrial and social organization (ibid: 111).

To summarize the argument: Once the first industrial nation had been born it provided the model to imitate for the rest of Europe (as well as her North American replica). Not to imitate would mean permanent dependence on the 'workshop of the world' and danger to the other nation state projects. This basic dilemma was to be repeated more generally in the relation between the West and the decolonized world. In order to develop it was deemed necessary for the 'new nations' to imitate the Western model – it was a 'modernization imperative' (Nayar 1972).

The Soviet state was also consolidated by war against both internal and external enemies. During the October Revolution Lenin said: 'Either perish or overtake and outstrip the advanced capitalist countries'. In the early 1930s Stalin echoed: 'We are fifty or a hundred years behind the advanced countries. We must make good this distance in ten years. Either we do it or they crush us.' (Both quotes from Holloway 1981: 9). This was the modernization imperative as it appeared to the Russian revolutionaries.[16] The emergence of the Soviet system was the ultimate result. This system is inseparable from the present division of the world into two hostile blocs. Both superpowers in fact define security in terms of bloc stability, and not merely in terms of defence of national territories.

The European model of development was violent not only in certain crucial historical phases, but remains inherently unstable in the way it moulds the international structure of trade. It is quite natural that since the politicomilitary impulse leads to the reproduction of similar structures of production, and because (for reasons of economies of scale) the strategic sectors are designed also for export, a surplus capacity is created. From this ensuing economic rivalry there is a logical, if not necessary, step to trade wars and military wars.

As a matter of fact protectionism rather than free trade has been the predominant economic policy for those emerging states that have been allowed to have their own say. International regimes characterized by free trade necessitate a stable economic order in which one dominant or hegemonic state (Great Britain and the USA are two historical examples) decides upon and guarantees the rules of the game. In consequence the weakening of hegemonic power implies more uncertainty and more conflict, perhaps even war.

2.1.2 Mainstream and counterpoint

Methodologically it is a great problem to pinpoint what characterizes one's own society. Therefore a stranger has certain advantages when it comes to analysing the distinctive quality of a particular society (Srinivas 1969). What is too obvious to be noticed by its own members can be illuminated by a process of cultural translation. Waiting for North–South intellectual exchange to become more symmetric and for Southern anthropologists to flood the Northern metropoles, we have to be content with contributions which attempt to transcend the Eurocentric tradition and look upon Western development, as it were, from outside.

For analytical purposes we make the following two distinctions:

- First, between the general level of Western cosmology or belief-system on the one hand, and its more concrete manifestations in contextually and historically specific development strategies on the other.
- Second, between what may be called the 'mainstream' of Western development thinking and its 'counterpoint'.

Development is one of the oldest and most powerful of all Western ideas. The central element of this perspective is the metaphor of growth, i.e. growth as manifested in the organism (Nisbet 1969: 8). Development, in accordance with this metaphor, is conceived as organic, immanent, directional, cumulative, irreversible and purposive. Furthermore, it implies structural differentiation and increasing complexity.

Certainly, the emphasis within the Western growth perspective shifted as new elements were added during the history of European civilization. Thus, the emergence of capitalism, the rise of the bourgeoisie as a dominant class and the industrial revolution all gave a distinct shape to Western developmentalism. This is the reason some writers stress the intellectual movements in seventeenth-century Europe as the cradle of Western development thinking.

The most significant shift in emphasis was the identification of growth with the idea of progress (Bury 1955). This was according to Bury a novel emergent in the Western mind, since growth in Greek and Roman civilization basically was thought of as a cyclical process. Medieval authorities conceived growth in terms of degeneration, decay and with a certain sense of doom. The modern idea of progress, in contrast to both these views, implied that 'civilization has moved, is moving, and will move in a desirable direction'. [17] The most explicit expression of this conception may be found in the works of Condorcet, Saint-Simon, Comte, Spencer and Marx. Certainly, different dimensions of 'progress' were emphasized by different social thinkers. To Condorcet and Comte knowledge was in focus, whereas Marx, in contrast, stressed the progressive movement of the productive forces. [18]

The founders of anthropology (i.e. 'armchair anthropology'), who dealt with 'other cultures' more explicitly, used the evolutionist framework for taxonomic purposes in what was known as the Comparative Method. This 'method' was not really comparative in the modern sense of the word, but simply an expression of the 'belief that the recent history of the West could be taken as evidence of the direction in which mankind as a whole would move and, flowing from this, should move' (Nisbet 1969: 191). The comparison was between 'backward' and 'advanced', 'barbarian' and 'civilized' and 'traditional' and 'modern'. The latter distinction was, as we shall see, to play an important role in development theory. Among the evolutionists associated with the Comparative Method the best-known is probably Lewis Morgan, simply due to his influence on Marx and Engels. The latter drew heavily on Morgan's *Ancient Society* in writing his *The Origins of Family, Private Property and the State*.

The Marxist approach to development was thus not unaffected by the evolutionary idea of progress prevalent in the nineteenth century. [19] For Marx development was first of all development of capitalism. New, higher relations of production (e.g. socialism) could in his view not appear 'before the material conditions of their existence have matured in the womb of the old society'. The notion of underdevelopment did therefore not exist in the classic Marxist system, where 'the industrially more developed country showed to the less developed the image of its own future'. If backward countries suffered from anything, it was this incompleteness of capitalist development. Marx's analysis of India and China in terms of an Asiatic mode of production suggested that these countries were unable to develop without the aid of colonialism. Marx's and Engels's views on Latin America were not much different. The best that for example a country like Mexico could hope for was to be dragged into history by the more advanced United States (Aguilar 1968: 66–7).

The Marxist outlook was of course modified as society changed and capitalism went into its monopoly stage, but the inevitability and essentially progressive nature of capitalist development was never abandoned. In his *Development of Capitalism in Russia* (1899) Lenin echoed Marx's and Engels's later views that the precapitalist social structures of Russia were doomed, and that capitalist development must be regarded as progressive in spite of its social costs. However, the special characteristics of Russian society created atypical conditions for its development. Marxism did not come closer than this observation in developing a theory of underdevelopment (Palma 1978). As has been argued by Foster-Carter, the 'absence of relationality' is the basic reason for this (Foster-Carter 1976b). Another way of putting it is by reference to the role of the 'endogenous paradigm' in Western intellectual history (Smith 1973). It was this bias that was tackled by the dependency school.

So far we have discussed the predominant mode of Western development thinking. Opposed to and dialectically related to this mainstream paradigm, there has been a counterpoint, articulating diverse interests and expressed in varying historical contexts, but essentially arguing for the inherent superiority of small-scale, decentralized, ecologically sound, human and stable models of societal development. Although an old ideal (as old as industrialism) one of its most well-known contemporary formulations is the one given by Schumacher: 'Small is beautiful'. This ideal is also sometimes referred to as 'neo-populism' (Kitching 1982).

The counterpoint may be traced back to premodern structures, but should not be interpreted simply as nostalgic romanticism, even if this is one of its several manifestations. We may find equally typical expressions in anarchism, utopian socialism, populism and other ideologies articulating protests against modernity. The counterpoint protest has more and more become confined to the ideological level as the modern complex was institutionalized in structures such as the state and the bureaucracy, the industrial system, the urban system, the professional elite, the techno-scientific system, the military-industrial complex, etc. These structures dominate the industrial societies whether they are socialist or capitalist, and the vested interests in them are of course immense. What the contemporary counterpoint position can draw strength from is therefore a gradual weakening of the modern complex as its maintenance costs increase, and the economic growth it is supposed to guarantee fails to come about.

The Counterpoint is hard to describe and as soon as one tries it tends to get dissolved in abstractions or trivialities. However, taking our point of departure in a negation of the modern complex, a society organized according to counterpoint ideals would be physiocratic, in the sense that the earth and the natural resources constitute the ultimate preconditions for human existence, ultra-democratic, in the sense that people exercise control over their own situation, and structurally undifferentiated, in the sense that the division of labour is within man rather than among men. In Chapter 5 we will return to these ideals in terms of development theory and look into some of their concrete manifestations in more detail.

2.1.3 Mainstream manifestations

Mainstream development thinking can be analysed along a continuum running between two ideological antipoles, socialism versus capitalism. Along this horizontal axis we can situate the concrete development strategies that have emerged in the course of European economic history. Thus, much of the political debate in the West has been concerned with State versus Market, and the relative merits of these antagonistic institutions in the context of economic development. On the other hand the mainstream – counterpoint polarization, discussed above, is an expression of quite a different dimension, which only recently has become politically articulated (see Fig. 2.1).

Along the horizontal dimension it is possible to identify several more or less distinct development strategies within the mainstream Western tradition: the liberal model, the state capitalist strategy, the Soviet model, Keynesianism. The liberal market-oriented model has been something of a norm but historically not the most important strategy. As more and more countries during the nineteenth century tried to catch up with Britain, the state assumed a crucial development function (Senghaas 1985). In the mid-twentieth century capitalism Keynesianism, expressing the increased role of state power, became predominant. The other

Fig. 2.1 State–market and Mainstream–Counterpoint polarization

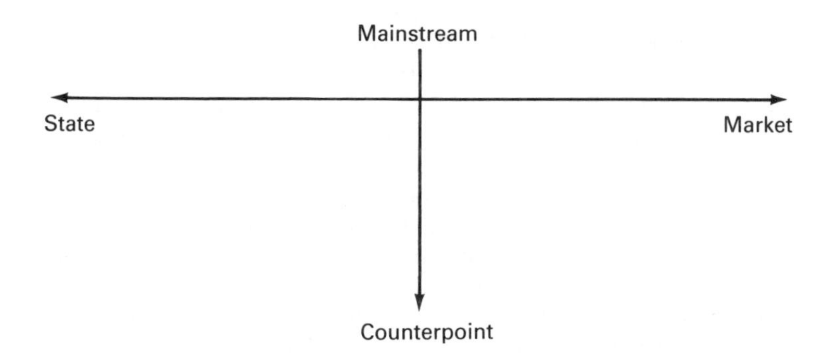

development strategies are varieties of the basic paradigm, expressing different historical possibilities and constraints. They differ mainly with regard to means (i.e. the relative role of state and market) but as far as the ends (the Western conception of modernity) are concerned, they are all basically similar. The differences as regards means can largely be explained by the specific circumstances in which the strategies emerged. Let us briefly consider these circumstances, starting with the more market-oriented strategies.

The liberal model is unique and constitutes, in generalized form, the English development experience in the era of industrial revolution. The model implied reliance on market forces, gradual industrialization, starting with light industries, a sufficient level of private investment derived from high profits and a low level of wages, and a stress on technological advancement, which necessitated capital accumulation and expanding markets. The repetition of this development path became increasingly difficult with the emergence of a capitalist world economy and the structured division of labour between the participating countries. To become developed like England was still the goal, but the means had to be reconsidered in the light of her superiority as the 'workshop of the world'. Significantly, the ideas of development contained in classical political economy were by the less developed countries considered as a development ideology, and, what is more, as a false one. These industrializing countries (the nineteenth century NICs or Newly Industrializing Countries) were instead protectionist and relied on state power for preparing themselves for the takeoff.

The state capitalist strategy thus belongs to an early phase of industrial development in continental Europe. It was typically an attempt at enforced development in primarily agrarian economies, the prime example being the policies of Count Witte in tsarist

Russia. Witte was influenced by the German economist Friedrich List who saw industrialization as necessary for nationalist and security reasons (the 'modernization imperative'). However, List never abandoned his basically liberal outlook (von Laue 1963). As soon as state-fostered industrialization had been carried through it was, according to List, beneficial to participate in the international division of labour, but not before. The same function of modernization as a nationalistic reaction to a threat from industrially more advanced countries (rather than being endogenous and imminent) applies not only to Germany and Russia but also to non-Western countries, like Japan after the Meiji restoration.[20] In Russia the stubborn populist resistance against Witte's industrialization programme was substantially weakened after the humiliating defeat in the Russo–Japanese war (Mendel 1961).

Keynesianism, in contrast to the state capitalist strategy, is a manifestation of mature capitalism. Its departure from the basic liberal model consisted in granting the state a responsibility for the stability and continuous growth of capitalist systems. It is not altogether correct to speak about Keynesian theory as a development theory, but there were elements in the Keynesian theoretical revolution that could be incorporated in a more long-term growth strategy, characterized by state intervention and planned investment. This was, as will be discussed below, an important departure for the emergence of development economics.

Since the 1930s Keynesianism has been the dominant development ideology in the industrialized capitalist world, particularly in countries with strong social-democratic parties, eager to replace their original socialist doctrine with the supposedly more 'scientific' Keynesianism. In the aftermath of the world economic crisis in the 1970s, particularly in view of the inexplicable (in terms of Keynesian economic theory) phenomenon of 'stagflation', the ideology of Keynesianism suffered many setbacks and lost influence at least as a national strategy.

Today state-fostered growth is consequently less popular and Keynesian orthodoxy challenged by a neoliberal wave in the industrial countries, looking for ways out of the impasse. All over the industrialized capitalist world, 'liberalizing' economic experiments have been started, in some countries with great enthusiasm, in others more reluctantly. It is only natural that the major alternative within the predominant paradigm – i.e. the market – is tried, when the old medicine has lost its curative power.

Let us now turn to the Western experience of a more radical state-oriented strategy. The Soviet model was to a large extent a continuation of the state capitalist policy of prerevolutionary Russia, although the ideological inspiration and the political context differed. In the famous economic policy debate of the 1920s

(Erlich 1960) several options implying different development paths were discussed.

Ultimately Stalin responded to the modernization imperative, dramatized by the military threat from Germany. The outcome was primitive socialist accumulation enforced by state-intervention and military force. Stalin's five-year plans contained what later was to be known as the Soviet model. It can be summarized as follows:

- The pattern of economic development was designed in five-year plans of a mandatory nature.
- The process of resource mobilization implied a transfer of resources from agriculture to industry.
- The agricultural sector was collectivized in order to facilitate control over the process of resource mobilization.
- As for industrial investments, priority was given to heavy industry.
- The industrial development was characterized by large-scale and technologically advanced methods.

In accordance with our distinction between theory and ideology the Soviet model based itself on theory but was transformed to ideology when it was imposed on other countries as the proper development path. Granted that Stalinism, upon which modern Soviet state and society were built, can be seen as a variety of Marxism corresponding to the Western tradition of materialism and growthmanship, there should be no controversy about looking at both the capitalist and the socialist development strategies as expressions of the Western paradigm of development. The development projects of the Soviet model were part of a substitution process, where alternative means were made use of in order to reach similar ends as in Western Europe (Gerschenkron 1962). The modifications of the economic policy in the post-Stalin era have only further underlined this, as we shall come back to in Chapter 6.

What is striking when one finally compares the Western development strategies is the important role given to the state in spite of prevailing orthodoxies. The state-capitalist strategy made the state responsible for capital formation and investment in strategic areas, where spontaneous demand was lacking. In the Keynesian strategy the controlling agency in the process of economic growth has been the state rather than the market. In the Soviet model the state completely replaced the market mechanism. Later experiences with profits and prices have so far not substantially altered this picture. As will be discussed below the more state-oriented strategies turned out to be more attractive for the 'modern elite' in the Third World, since the elite, with very few exceptions, has fostered an antimarket bias. In practice the three strategies merged into the development ideology of the ruling class in most underdeveloped countries. Whether the revival of the long forgotten ideal model of liberal capitalist development will have the same impact on the Third World remains to be seen.

2.2 The rise and decline of development economics

It is generally agreed that development economics as a distinct sub-discipline (to the extent that there is agreement on it being distinct) emerged only after the Second World War. As has been emphasized by the late Gunnar Myrdal the major recastings of economic thought have always been responses to changing political conditions and opportunities (Myrdal 1968: 9). What specific political conditions gave birth to development economics?

According to Myrdal they were:

- The rapid liquidation of the colonial power structure.
- The craving for development in the underdeveloped countries themselves.
- The international tensions, culminating in the cold war, that have made the fate of the underdeveloped countries a matter of foreign policy concern.

Myrdal also noted that 'we economists are riding the crest of the wave' (op. cit: 8). In what follows we shall trace the emergence of development economics in the context of the political concerns outlined by Myrdal. Secondly we deal with some of the pioneers in the establishment of the discipline, and thirdly what many observers see as the decline of development economics, or 'the counter-revolution'.

2.2.1 Development rediscovered

After the Second World War the largely destroyed Europe was herself in need of development. Instrumental in this development – or rather reconstruction – was the massive aid programme from the US, the Marshall Aid. This programme had the double purpose to get the world economy working (in accordance with the Bretton Woods system) and to contain communism.

From the reconstruction of Europe there was a straight line to the problem of development in the new nations, known by many names: backward, underdeveloped, emergent, poor, less developed, developing, etc. 'Colonial economics' became 'development economics'. The early optimism in this field must be seen against the background of successful European reconstruction. As Preston points out (Preston 1982) there were three pillars behind the emerging development theory: growth, planning and aid; and the overall conceptualization was based on the idea of manipulative intervention. Later elaborations were related to the experiences of implementing this idea in various Third World contexts. The newness of these experiences and the many setbacks associated with them gradually gave rise to the claim that development economics was not only a distinct academic field but also in need of a special institutional set-up. One reason was the gradual discovery of what became known as 'noneconomic factors'.

In an article with the title 'The Rise and Decline of Development Economics' Albert Hirschman (1981) explains the fast emergence of development economics by the combination of two methodological and theoretical positions:

- Rejection of the monoeconomics' claim, which implied a need for a separate theoretical structure.
- Assertion of the existence of mutual benefits between rich and poor countries, which made development economics relevant from the aid donor's point of view and created a demand for development economists.

These positions contrast with two orthodoxies – neoclassicism and Marxism – and thus provided development economics with its claim for originality without being unacceptably radical. The originality was provided by a series of remarkable pioneers to be presented below.

According to Hirschman, development economics took advantage of the discredit orthodox economics had fallen into as a result of the depression of the thirties and the Keynesian revolution which established a conception of two kinds of economics: 'The Keynesian step from one to two economics was crucial: the ice of monoeconomics had been broken and the idea that there might be yet another economics had instant credibility' (Hirschman 1981).

As for the mutual benefit claim, soon to be rejected by radical underdevelopment theory, it inspired confidence in the manageability of the development enterprise and placed it on the agenda of policy makers all over the world. The erosion of this position by later developments in theory and in the real world is also Hirschman's explanation of the more recent decline of development economics, already referred to in the previous chapter.

It is quite natural that the original recipe for development given by the developed countries should emanate from their own experiences and prejudices. Above we tried to outline what these experiences and prejudices were in terms of development ideology. Our purpose here is to look into the academic form of Western development thinking, primarily in mainstream economics, leaving the rest of the social sciences, as well as earlier Marxist contributions, to the last section.

The dominant approach in economics of development in the late 1940s has been summarized by Nugent and Yotopoulos as follows:

> . . . change is gradual, marginalist, nondisruptive, equilibrating, and largely painless. Incentives are the bedrock of economic growth. Once initiated, growth becomes automatic and all-pervasive, spreading among nations and trickling down among classes so that everybody benefits from the process.
> (Nugent and Yotopoulos 1979: 542)

The neoclassical paradigm can be studied in its most typical form in the theory of international trade where it was postulated that free trade will spread the benefits of development across the world through the kind of specialization and division of labour enforced by competition. In the neoclassical world underdevelopment does not exist except as a lack of entrepreneurial spirit.

This paradigm thus was an optimistic one, rather different from the 'dismal' view of the classical economists and even the Schumpeterian and Keynesian traditions, succinctly summarized by Hans Singer:

> Ricardo's great worry and gravedigger of progress was, of course, the Law of Diminishing Returns in agriculture. Malthus' great worry and gravedigger was population. Karl Marx' gravedigger was collapse of markets and insufficiency of purchasing power. Schumpeter's chief worry and gravedigger was the undermining of the entrepreneurial spirit, whether from public hostility or from a 'Buddenbrooks complex', under which the sons and grandsons of successful entrepreneurs become poets or collectors of art or give their money away. In Keynes' view, the villain of the piece and gravedigger of progress was the 'falling marginal efficiency schedule of capital': as capital accumulation proceeded, new investment opportunities were gradually used up; and as the rate of interest could not, for various reasons, continue to fall in step with the falling marginal efficiency of capital investment, capital accumulation and progress would come to a stop. (Singer 1978: 2)

On the whole the Keynesian interventionist line of thinking became dominant in the field of development economics, whereas the neoclassical view was relegated to the backbenches. Thus there were several theoretical inputs in the new science of economic development but the main message was that development necessitated plans, written by economists, and strong, active governments to implement them. Development was an art of large-scale social engineering.

We have earlier touched upon the question why this 'received knowledge' was largely accepted by the ruling elites in the Third World. The state capitalist strategy was basically a response to the modernization imperative ('industrialize or perish'), the Soviet strategy could draw prestige from the success of Soviet industrialization, while the Keynesian strategy ensured the economists the position as the high priests of growth. It is therefore only logical that comparatively few countries in the third World were attracted by the economic recommendations implied in M. Friedman's neo-liberalism, not so much because the social consequences were considered problematic, but rather because they were less consistent with the existing power structures in the Third World.[21]

It has been argued that these power structures implied an urban bias in planning and development (Lipton 1977). To this urban bias I would add an antimarket bias which may not be equally all-pervading, but sufficiently strong in most parts of the Third World to favour that part of the received knowledge which emphasized the importance of central planning and the role of the state. Typically, indigenous capitalists were not expected to contribute to the releasing of productive forces; to industrialize and modernize. The social group that was entrusted this historical task was instead the modern Western-educated elite, often recruited from traditional groups with a marked bias against the commercial way of life. One obvious case in point is the Brahmin caste of India, but also in Africa the 'modernizers' tended to be lawyer-politicians rather than merchants. In Latin America more hope was perhaps pinned on the national bourgeoisie, at least until it became unmasked by the *dependentistas* as a 'lumpen bourgeoisie', tied to foreign interests and regarded as carriers of underdevelopment rather than development.

It is thus somewhat paradoxical, but not inexplicable, that of those development strategies which have emerged from Western historical experiences, the one that in retrospect appears as least relevant – the Keynesian – came to be of most immediate importance for the birth of development theory. The reason was of course a coincidence in time (the late 1940s) between Keynesian theory as a predominant school in economics and the growing interest for 'backward areas', by which name the underdeveloped world was known in those days. The character of the modern elite and its close relations to the state made strategies which emphasized the role of state power more attractive than purely liberal alternatives. The Western model was thus not in any way forced upon the developing countries. It was fully consistent with the power structures in those countries.

Keynes's theoretical contribution implied that macro-economic problems returned as a key issue in economics. In times of depression the problem was to employ existent, but poorly utilized factors of production. Keynes held that the aggregate demand, and its various components, e.g. consumption and investment, were of strategic importance. An increase in expenditures which, in turn, multiplied aggregate demand, would eventually lead to an increase in the level of economic activity and a decrease in unemployment. Thus, Keynes was primarily interested in short-term problems of stabilization. Since the problems of under-development are radically different from those of the depression he did not directly contribute to the theory of economic development: he was not very interested in long-term economics and held the opinion that countries which had reached a certain income level should stop being too concerned about economic growth (Singer 1978: 24). Therefore it is somewhat unfair to associate Keynes's name with growthmanship, but this is what his theoretical framework later has been used for, in spite of the rather specific circumstances which gave birth to it.

In the Harrod–Domar model, derived from Keynesian theory, each increase in output provides the basis for further growth because part of the increased output is reinvested. At higher income levels the marginal propensity to save is higher and therefore economic growth, once the process has started, will be self-sustaining. The model has been elaborated to account for various levels of population increase and different capital–output ratios. Still it was fairly simplistic in its stress on capital accumulation as the main factor in the process of economic growth. The further 'development of development thinking' consisted in elaborations of this basic model, as well as the parallel body of thought, concerning the rationale of trade between nations, which was derived from classical and neoclassical rather than Keynesian theory. Thus one line of theorizing concerned the endogenous process of growth, another trade relationships. There was in fact very little connection between the two.

2.2.2 Pioneers in development

The title of this subsection is taken from a series of World Bank public lectures devoted to ten pioneers in development economics: P. T. Bauer, C. Clark, A. O. Hirschman, A. Lewis, G. Myrdal, R. Prebisch, P. N. Rosenstein-Rodan, W. W. Rostow and J. Tinbergen.[22] It is interesting to note that a majority of them originate from Europe (two from Eastern Europe), whereas two come from the Third World and one from the United States. Most of them were more or less interventionistic, i.e. they believed in planning and were mainly concerned with the mobilization of domestic resources. Development was seen as a national process. Thus a cleavage developed between a Keynesian, 'nationalist' school, and a more orthodox, neoclassical tradition, stressing the importance of trade and comparative advantages in the development process.[23]

Most of the pioneers held that neoclassical theory due to its preoccupation with microeconomics had little to offer the theory of development. However, a few defended the conventional theory of trade, ultimately developed from the classic arguments about the rationalizing effects of large-scale production and widening markets, and the efficiency of market exchange. Foreign trade was seen as an 'engine of growth', especially for the late-comers in development. According to Jacob Viner (1953), the major benefits to those countries were:

- reduction in trade barriers;
- free international movements of capital;
- diffusion of technical knowledge and skills.

This line of thought was developed alongside, and often in opposition to, the various theories dealing with internal growth factors. Many of these actually presupposed a 'closed economy', an assumption that the trade theorists considered both unrealistic and irrelevant.

Proponents of balanced growth, as well as those arguing for unbalanced growth, stressed the importance of complementarities, the first school at a specific point in time, the latter over time. Trade theorists, however, argued that the most important complementarity was between national economies cooperating within an international division of labour. The basic argument is the theory of comparative advantages, stating that a country may raise its level of consumption above what would have been possible in a state of autarchy by specializing in the production and export of commodities which, relatively, have the lowest costs of production. The world as a whole, as well as each single country (including the less developed ones), obtains more goods, at a constant level of factor input, through an international division of labour in which all nations specialize in and export those commodities which they can produce most effectively. In a country with a relatively good supply of labour, the cost of labour tends to be low as compared with other countries. Thus, the country in question should have a comparative advantage in the production of commodities which use more labour than capital. It should therefore export labour-intensive commodities and import capital-intensive ones from countries with relatively ample supplies of capital. The theory furthermore stated that free trade would lead to factor price equalization between the countries. For example the wage differences between the developed and the less developed countries would be reduced, which, in turn, would lead to a more equal distribution of income in the world.

It should, however, be noted that very few Western economists could fall back on much concrete experience of development problems in 'backward areas'. Those who had such experiences soon began to question the relevance of conventional economics, starting with international trade. In 1949 Hans Singer delivered a paper to the American Economic Association where he questioned the assumption that participation in international trade necessarily made all parties concerned better off.[24] He pointed out that the textbook argument never had been really accepted by the more articulate economists in the underdeveloped countries. He also gave a number of reasons why the distribution of gains from trade tended to become uneven. The most important, known as the Singer–Prebisch thesis, was that the underdeveloped countries producing primary commodities were subjected to deteriorating terms of trade *vis-à-vis* the industrialized nations. Contrary to what the proponents of free trade had suggested, the underdeveloped countries should therefore not produce and export commodities in which they had comparative advantages, but instead attempt to change the whole structure of comparative advantages. The way to achieve this, Singer then believed, was to invest in industrialization.

Gunnar Myrdal came rather close to a theory of underdevelopment when he took one step further and pointed out that the so-called 'backwash' effects from trade might lead to severe misallocations and deepening stagnation (Myrdal 1956). His principle of cumulative causation, which could be positive as well as negative, was spelled out in explicit refutation of equilibrium analysis, which in Myrdal's opinion was completely irrelevant in the study of development (Myrdal 1957). Myrdal's discussion did not only refer to international trade but was also meant to illuminate the problem of inequality between regions both internationally and within nations. Myrdal's criticism of the conventional neoclassical paradigm admittedly was destructive to it but did not really point out the new directions which development theory should follow.[25]

Development was, in post-Keynesian as well as Marxian economics, primarily seen as a process of capital formation determined by the level of investment. This postulate was carried over to development economics. It was emphasized in the first UN report devoted to development problems, *Measures for the economic development of under-developed countries* (1951). According to one of the authors and a true pioneer in development economics, the West Indian economist W. A. Lewis:

> The central problem in the theory of economic development is to understand the process by which a community which was previously saving and investing 4 or 5 per cent of its national income or less, converts itself into an economy where voluntary saving is running at about 12 to 15 per cent of national income or more. (Lewis 1954: 155)

This by now classical statement gives a good idea about the general perspective in which the bulk of development literature of the

1950s was produced. The basic problem in the underdeveloped countries was how to break loose from the fetters that prevented them from marching along the growth path, mathematically symbolized in the Harrod–Domar model. Before the famous aeronautic metaphor of 'taking off' into 'self-sustained growth' carried the day (more about Rostow's stage theory in the next section), there were several models that tried to illuminate the various 'traps' and 'vicious circles' that the underdeveloped countries were struggling with. For example population growth (the dimensions of which were revealed by many population censuses published in the early 1950s) was seen as an important restraint (a population trap) on economic growth by several authors, e.g. H. Leibenstein (1957). Since a moderate level of growth was eaten up by the population increase, a 'critical minimum effort', normally conceived of as some external stimulant, was needed.

This idea of a trap, which certainly bears resemblances to Keynes's low-level equilibrium, provided a rationale for foreign aid as the key to economic development. The 'critical minimum' could of course never be defined beforehand, so if development did not occur the blame could be put on insufficient foreign aid. Structural factors, internal or external, were not considered. The basic idea was growth as a unilinear process which would continue eternally once momentum was gained.

To illustrate this idea W. A. Lewis in an early article dealing specifically with West Indian industrialization used a snowball metaphor:

> . . . Once the snowball starts to move downhill, it will move of its own momentum, and will get bigger and bigger as it goes along . . . You have, as it were, to begin by rolling your snowball up the mountain. Once you get it there, the rest is easy, but you cannot get it there without first making an initial effort. (Lewis 1950: 36)

In a similar vein Paul Rosenstein-Rodan (1943 and 1961) and Ragnar Nurkse (1953) made the observations that in an underdeveloped economy, characterized by a 'vicious circle of poverty', the investment programme must be both massive and balanced, for growth to be sustained. These authors were less optimistic about the role of foreign aid and put emphasis on domestic saving and the role of the state. The basic evolutionary framework was nevertheless the same. In the absence of a big push and balanced growth, the Harrod–Domar model did not describe a process of growth but of infinite stagnation or, to return to the Keynesian language, a permanent low-level equilibrium.

As empirical knowledge about the 'backward areas' grew, the fundamental difference between the phenomenon of underdevelopment and a situation of low-level equilibrium in an advanced economy became increasingly obvious. A big push could for example not be expected simply because these countries were underdeveloped, not only in terms of investment inputs but also in terms of decision-making, entrepreneurship and administrative efficiency. In this context A. Hirschman (1958) proposed a strategy of 'unbalanced growth' which admittedly was more realistic with regard to different bottlenecks. Bottlenecks were here seen as inherent in the process of development which therefore by necessity had to be unbalanced. Hirschman and other spokesmen for unbalanced growth conceived these bottlenecks as challenges to be overcome by, but also as stimulants for, growth. This vision of development as a 'chain of disequilibria' was paralleled in regional analyses in the works of F. Perroux (1950, 1971) where it was pointed out that growth did not appear everywhere at the same time but manifested itself in 'poles of growth'. These approaches were far more realistic since they were disentangled from the neoclassical obsession with equilibrium and even touched upon the idea of polarized growth (later to turn up in dependency analysis). But still the major problem was to remove obstacles. Behind the ideas of unbalanced growth and growth poles it is thus easy to recognize the conception of growth as a more or less natural and automatic but occasionally disturbed or interrupted process.

One markedly non-Western phenomenon, characteristic of most situations of underdevelopment, which soon caught the attention of the pioneers in development theory (still identical with development economics) was the 'dualistic' nature of underdeveloped economies: the coexistence of an advanced or modern sector with a backward or traditional sector. This provided the background to one of the more influential theories of development formulated in the 1950s, namely W. A. Lewis's strategy of development in an economy characterized by 'unlimited supply of labour' (Lewis 1954). In his theory Lewis combined ideas derived from classical economics with his own experiences of the West Indian economies, where the number of unemployed was high and the capacity of agriculture to absorb an increasing labour force had been exhausted. The way out of this impasse was, according to Lewis, industrialization, financed by foreign capital (Industrializing by Invitation, as Lewis's critics termed it). The process of industrialization would continue with labour cost at a fixed level so long as there existed labour with zero marginal productivity in the traditional sector.

The idea of dualism was the central theme also in many later contributions to development theory that transcended the strictly economic framework. It may appear as if this idea signified the abandoning of the unilinear, evolutionistic framework, but this is really not the case. Rather the two sectors, the traditional and the modern, were conceived of as two stages of development, coexisting in time, and in due course the differences between them were to disappear because of the natural urge toward equilibrium.

The barriers to development were to be found in the traditional sector, and as theorizing about development continued into the 1960s, these barriers came to be identified with various sociological, psychological and political factors often referred to as 'noneconomic factors' by development economists. W. A. Lewis, whom we have referred to often, wrote the first standard text in a broad, truly classic tradition (Lewis 1955). The 1960s saw the interdisciplinary broadening of development theory. The transition from a traditional state to modern economic growth presupposed changes in attitudes, sometimes defined as 'empathy' (Lerner 1962), sometimes as 'need for achievement' (McClelland 1962), as well as in social and political institutions (Hoselitz 1960, Apter 1965, Myrdal 1968). A lot of ground had thus been covered since the Harrod–Domar model first began to be applied to the underdeveloped regions. However, the core premise of the emerging paradigm was still the simple growth equation.

In a classic article entitled 'The limitations of the special case' Dudley Seers raised the awkward argument that conventional economic theory was valid only for Western industrial capitalism (1963). He proposed the 'modest but revolutionary slogan' that 'economics is the study of economies', which implied that different types of economies were in need of different types of theoretical frameworks. As far as developing economies were concerned, Seers felt that the Latin American structuralists were on the right track, and in this context he also mentioned the works of Myrdal, Singer, Nurkse and Chenery as harbingers of a new 'paradigm', although this concept was not in use among social scientists at that time. In spite of this, Seers's argument had a distinct 'Kuhnian' flavour. The following quotation for example is a remarkably apt description of a 'paradigm crisis':

> Economics seems very slow in adapting itself to the requirements of the main task of the day – the elimination of acute poverty in Africa, Asia and Latin America – just as the previous generation of economists failed to cope realistically with economic fluctuations until after the depression had brought politically catastrophic results. (Seers 1963: 77)

In the intellectual process that led to the Keynesian revolution Seers distinguished three phases: the Hobson phase (the discovery of underconsumption), the Kahn phase (the idea of the multiplicator) and the Keynes phase (the synthesis). However, the road towards a relevant development theory had not yet been discovered:

> We are not even in a position to judge what has to be demolished of the old doctrine, or what can be saved and adapted for further use. This is still somewhat in the 'Hobson phase', or early in the 'Kahn phase', of development economics. The 'Kahn article' may not yet have been published. A characteristic of this article is, of course, that nobody recognizes it as much at the time; only in the 'Keynes phase' can its historical significance be appreciated. (ibid: 79)

Seers's interesting question concerning the harbingers cannot be answered today, two decades later. We will meet several 'Kahn articles' in this survey but not the 'Kahn article', and certainly not Keynes.

The early discussion on development had an optimistic tone which is hard to explain today. It was a manifestation of the dynamic growth experienced by the industrialized countries themselves, supported by a philosophic tradition which looked upon growth as natural and therefore more or less inevitable. But one could quicken the speed. The simple formula was: just find out the incremental capital–output ratio and the desired rate of growth. Then you can (after due consideration of the rate of population growth) arrive at the appropriate level of investment. Growth was thus seen mainly as a function of investment and very few doubted that a process of economic growth through a series of 'stages' ultimately would benefit the whole nation. The 1960s were proudly given the name of the First Development Decade in expectation of what was to come. Those were truly innocent years. As we saw in the first chapter many theorists today look back on them with a certain embarrassment.

In fact the record of development since the 1950s is not so impressive if development is defined in a more comprehensive way than simply in terms of growth, as became more common in the 1970s. In their study *Economic growth and social equity in 'developing countries'*, dealing with the social effects of growth, Adelman and Morris (1973: viii) concluded: 'Our result proved to be at variance with our preconceptions.'

A similar statement was made by H. Chenery in the IDS/World Bank sponsored study *Redistribution with growth*: 'It is now clear that more than a decade of rapid growth in underdeveloped countries has been of little or no benefit to perhaps a third of their population.' (Chenery *et. al*. 1974: xiii.)

These two books are good examples of the renewed interest in the connection between economic growth and income distribution in the 1970s. The reasons for this emphasis were empirical findings on the poverty complex: marginalization, mass unemployment and recurrent starvation crises. All this contributed to political tensions and upheavals in many countries. Politics was knocking at the door. The green revolution was only one experience of the 1960s which confirmed the universal observation, that what was taking place in many countries during the development decade was 'growth without development' – but with poverty.

The new strategy implied in the 'redistribution with growth' discussion was a modification of, rather than a clear break with, previous strategies. First of all, the analysis retained much of the optimistic flavour of earlier growth models in asserting that the benefits of growth, empirically, have a tendency to be concentrated to the early stages (Adelman and Morris even speak about 'pre-takeoff stages') but that 'further increases in concentration are by no means inevitable'.

Secondly, the social engineering approach to development was adhered to, as is shown in statements such as 'to deal with the problems of poverty groups we need to design overall programmes or policy packages rather than a set of isolated projects' (ibid: xi). This was the old recipe of balanced growth extended to cover social development as well. If balanced growth was a difficult endeavour, as so many critics pointed out, balanced economic and social development as a planning strategy appears rather utopian. To incorporate social objectives in a growth model is a theoretical-technical problem. To attack mass poverty in its concrete manifestations, on the other hand, is a political problem.

The political dimension, thirdly, was to some extent recognized in *Redistribution with growth* where a few chapters were devoted to 'political framework' and 'the scope for policy intervention'. It was correctly observed that 'a strategy involving the annual transfer of some two per cent of GNP from the rich to the poor for one or two decades would not be accepted readily by the rich' and that 'whether measures necessary to effect such a transfer are implemented by a government is primarily a question of who wields political power, what they perceive their interest to be, and how free they are to maneuver' (ibid: 52).

Here we shall not be concerned with the originality of these observations but rather take note of the fact that the economic approach to development in the mid-1970s had been stretched to include considerations of political power. By that time, however, a more 'orthodox' approach had already emerged.

2.2.3 The counter-revolution

Development economics of the structuralist type was never accepted by the leading proponents of economic orthodoxy and, as the new approaches had been tried and found wanting, this position was strengthened and experienced a revival in what with some justification could be called a fundamentalist form. The belief in the market mechanism now became unlimited and, consequently, the trust in government intervention grew thin. The fundamentalists held that efforts at comprehensive planning had proved to be failures and that liberalization of foreign trade regimes had shown to have positive effects on both growth and welfare (Little 1982). The new theoretical trend, emerging in the 1970s and taking a position of dominance during the 1980s, has aptly been termed the 'counter-revolution in development theory and policy'.[26]

Of course this counter-revolution in the theory and praxis must be understood against the background of the general anti-Keynesian neoliberal wave in the West and the rise of the market ideology in the East. Thus, the counter-revolution forms part of a changing political climate affecting both domestic economics and politics in the industrialized countries and Euro-South relations.

At a deeper level the counter-revolution reacts against and rejects the Western guilt-complex which lay behind the 'third-worldist' posture so characteristic of both reformist and radical underdevelopment theory. The poor countries are poor mainly because of mismanagement. As Toye notes, 'the magnificent vision of the 1980s is of a world developing its resources and capacities in response only to the ups and downs of relative prices and the self-imposed stasis of limited government' (Toye 1987a: viii). The counter-revolutionaries reject the 'limitations of the special case' and argue for the unity of theory as well as policy, i.e. the unity of the orthodox approach.[27]

Who are the counter-revolutionaries and what is the content of their ideas apart from the belief in the magic and miracles of the market? In particular, what kind of development strategy do they recommend to the poor countries? The proverbial old wine seems to have been poured into equally old bottles. Their arguments do not rest on any analysis of the broader social and political situation in Third World countries but a criticism of departures from the right economic path. Their aim is to restore confidence in the market mechanism and the free trade policy. They certainly do not lack arguments, drawing on all the mistakes, distortions, white elephants and examples of corruption which have been too visible in so-called development during the past two or three decades. They choose to disregard the criticism – still on the whole valid – which paved the way for the alternative approaches, and concentrate on the lack of visible results of these approaches.

Toye summarizes (op. cit: 48) the main components of the new vision of growth as follows:

- the benefits of markets and the danger that government action will negate these benefits;
- the relative unimportance of physical capital compared with human development policies;
- the distorting effect of government economic policies.

The development theorist who most persistently made these points over the years is P. T. Bauer. One of his earlier books in this genre (from 1972) was characteristically called 'Dissent on Development'.[28] Today these views are no longer those of dissenters. Ian Little in 1982 published a major work on economic development which summarizes both the critique of structuralist development economics and the new thinking. He sees domestic reforms in the direction of liberalization as the golden road: 'The nature of the international scene is important for development, but it is not important enough, and does not offer sufficient scope for constructive policy action, to justify the amount of attention it attracts relative to domestic matters.' Similarly the influential World Bank economist Anne O. Krueger (1986: 196) describes the accumulated knowledge in development economics in the following way: 'The change can be summed up as an increased recognition of the importance of markets and incentives – and of the limits of government and central planning.' Obviously the history of development economics is now being rewritten by the victors.

As earlier mentioned (see note 22) the World Bank organized a second series of public lectures on pioneers in development, focusing on the resurgence of neoclassical economics. This provided an opportunity for the former rebels to re-establish the neoclassical paradigm in development economics. Gottfried Haberler sees the rise of development economics as forming part of the decline of liberalism after the Great Depression (Meier 1987: 53). Consequently the decline of development economics coincides with the rise of neoliberalism. Hla Myint considers the neoclassical resurgence as 'the most noticeable development in the subject since the 1960s (op. cit: 107). Basing his argument on a distinction between acceptance of formal neoclassical economic theory and advocacy of free market and free trade policies, he is clearly in favour of the latter, while being somewhat ambivalent towards the former. However, on the illiberal interventionist phase in development economics there is no ambivalence. The same goes for its claim to represent a special branch of economics. Theodor Schultz repeats the point from his Nobel lecture that 'a major mistake of much new development economics has been the presumption that standard economic theory is inadequate for analyzing the economic behaviour in low income countries' (op. cit: 23 and Schultz 1980: 639).

In the following section we shall consider the early contributions from other social sciences to what later was to be called the modernization paradigm. It is only with the emergence of this multidisciplinary paradigm that we can speak of development theory as the more complex approach which was defined in the introduction. Development economics is today considered as economics applied to developing countries, in spite of the claims by many of its pioneers that this was a new and distinct branch of economics.

2.3 The modernization paradigm

In spite of the increasing complexity of development theory as it grew more interdisciplinary, it is possible to distinguish in the new approaches the basic evolutionary framework so characteristic of Western mainstream development thinking. Most contributions – whether economic, political, sociological, or psychological – were rooted in a basic paradigm, now most commonly referred to as 'the modernization paradigm'. Development was seen in an evolutionary perspective, and the state of underdevelopment defined in terms of observable economic, political, social and cultural differences between rich and poor nations. Development implied the bridging of these gaps by means of an imitative process, in which the less-developed countries gradually assumed the qualities of the industrialized nations. The task of analysing the qualities to be imitated was shared between economists, sociologists and political scientists, so that some specialized in economic structure, others in human attitudes, social institutions or political development.

The Marxist variety of modernization stresses qualitative leaps which a society is forced to make because of the dialectics of internal contradictions, expressed through class struggle. All transitions to new stages of development are by definition 'progress', since every 'mode of production' will exhaust its potential before being replaced by a 'higher' mode. This is what makes the Marxist theory of change another theory of modernization. We shall later discuss specifically the transition, or rather transitions, to socialism.

Modernization means different things to different people at different times. Therefore this great tradition is haunted by a certain confusion. The concept has been used in at least three senses: as an attribute of history, as a specific historical transitional process and as a certain development policy in Third World countries (Smith 1978: 61). It is the third meaning which is of most relevance in the context of development theory but the problem is (and that probably accounts for the appeal of the modernization perspective) that the three meanings are blurred. Modernization policies (implying a rationalization and effectivization of economic and social structures) are not only seen as elements of a development strategy (which may succeed or fail) but as the working out of universal historical forces (the first sense) which bear a strong resemblance to the transition from feudalism to capitalism in Western economic history (the second sense). Thus there are among the modernizers fundamentalists, who believe in development as a basically repetitive process, and less rigid proponents, who see modernization as one aspect of social change. It is the existence of the former approach, where development is an endogenous process, realizing the potential inherent in more or less embryonic form in all societies, depending on their level of social development, which makes it possible to talk about a paradigm. The paradigm can be summarized as follows:

- Development is a spontaneous, irreversible process inherent in every single society.
- Development implies structural differentiation and functional specialization.
- The process of development can be divided into distinct stages showing the level of development achieved by each society.
- Development can be stimulated by external competition or military threat and by internal measures that support modern sectors and modernize traditional sectors.

2.3.1 Evolution of capitalist modernity

We shall discuss some well-known theories from different social sciences within the modernization paradigm taking the gradualist approach. They have been chosen to illuminate both the central core and the borderlines of the paradigm, but also to show the broadening of the social science interest in development.

The sociology of development

Since the main contributions to modernization theory came from sociology it is appropriate to start with this discipline. There are many 'grand theorists' in this tradition but the central figure was Durkheim who saw in the division of labour and the postulate of structural differentiation the motive force of modern societies. It should, however, be noted that the classical theorists were mainly concerned with the transition from 'tradition' to 'modernity' in Western Europe, although the universal relevance of this scheme sometimes was implied.

It is also of importance that classical views usually were rather ambivalent *vis-à-vis* the process of modernization, as shown by Weber's concern with bureaucratization, Durkheim's concept of anomie and Marx's preoccupation with alienation. In modernization theory the negative pole of the polarity shifted to 'traditional society', whereas *Gesellschaft* came to be seen as desirable (Wolf 1982: 12).

A third relevant observation is that modernization theory, as part of the larger evolutionary tradition, of which it forms the most recent expression, conceives of social change as a basically endogenous process. We emphasize 'basically' since external variables often were called upon to explain how the process of modernization was triggered off, as in Marx's and Engels's famous statements of the function of colonialism to drag stagnant societies into history. However, the potential for modernization lay dormant in the societies concerned. For those most faithful to the paradigm, modernization was a universal process characteristic of human societies rather than a concrete historical process taking place in specific societies during specific periods.

The classical framework appears in modern form in Talcott Parsons's pattern variables: particularism–universalism, ascription–achievement and diffuseness-specificity (Parsons 1951). To complete this chain of influences Bert Hoselitz, economist by training, was the first to apply Parsons's pattern variables to the problem of development and underdevelopment:

> In attempting a theory of economic growth the main problem in relating social and cultural factors to economic variables is to determine how the social structure of a less developed country changes into that of an economically advanced country. (Hoselitz 1960: 17)

In this perspective society went through development or modernization as particularism, ascription and diffuseness were replaced by universalism, achievement and specificity. In practice, modernization thus was very much the same as Westernization, i.e. the underdeveloped country should imitate those institutions that were characteristic of the rich, Western countries. Hoselitz edited the influential journal *Economic Development and Cultural Change*, founded in 1952, in which the modernization paradigm took shape.

To give a deviating example from sociology we may refer to Barrington Moore's classical work *The Social Origins of Democracy and Dictatorship* (1966). Here the unilinear scheme so characteristic of the modernization paradigm was abandoned altogether and instead three historically distinct paths to modernization were identified: the classic bourgeois revolution (Britain), revolution from above (Germany) and revolution from below (Russia). Instead of ahistorical grand theory, the primary aim is to make sense of historical patterns. This is historical sociology liberated from the Parsonian framework but raising similar big questions (Skocpol 1984). There is no determinism in the analysis of the transition and no assumption of convergence after the transition. In accordance with the modernization paradigm, however, Moore's approach to historical development considered endogenous factors only.

Stages of economic growth

Probably the best known of the economic contributions within the tradition of modernization theory is that of Walt Rostow, who conceived of development as a number of stages linking a state of tradition with what Rostow called 'maturity'. This development was analysed primarily as an endogenous process (Rostow 1960).

Walt Rostow's doctrine which played an important political role during the late 1950s and 1960s was a typical expression of the Western development paradigm in its capitalist form. There were five stages through which all developing societies had to pass:

- the traditional society;
- the pre-takeoff society;
- takeoff;
- the road to maturity;
- the mass consumption society.

The economic prerequisites for a takeoff are created during the second stage, and many of the characteristics of the traditional society are then removed. Agricultural productivity increases rapidly, and a more effective infrastructure is created. Society also develops a new mentality, as well as a new class – the entrepreneurs. The third stage, the takeoff, is the most crucial for further development. It is during this period, covering a few decades, that the last obstacles to economic development are removed. The most characteristic sign of the takeoff stage is that the share of net investment and saving in national income rises from five per cent to ten per cent or more, resulting in a process of industrialization, where certain sectors assume a leading role. Modern technology is disseminated from the leading sectors while the economy moves towards the stages of maturity and mass consumption. According to Rostow, international relations do, in fact, speed up the process of development, but have little to do with underdevelopment. Rostow differed from the early development theoreticians by his much broader approach (he saw his theory as an alternative to the Marxist theory), but the key element in his thinking is nevertheless the process of capital formation.

It is interesting to compare this analysis with that of Alexander Gerschenkron, which differed in three important respects from Rostow's. In our view these differences demonstrate up to what degree one can depart from the basic premises of a paradigm without breaking it. First of all Gerschenkron's analysis (of the industrialization process of different countries in Europe) concerned a specific historical process, whereas Rostow's stages were supposed to be universally valid. Secondly, what appeared as preconditions for takeoff in Rostow's analysis here came out as results of the process of development. This process could be initiated in spite of lacking preconditions through what Gerschenkron called 'processes of substitution'. The most important factor was the state whose activities could compensate for lack of entrepreneurship, capital markets, etc. (cf. the discussion on the state-capitalist strategy above). Thirdly, the international context entered Gerschenkron's analysis as an important causational factor, which, through the possibility of substitution, made it possible for a country to take advantage of other countries' technology and thereby skip stages. In a key formulation Gerschenkron suggests that 'the industrial history of Europe appears not as a series of mere repetitions of the "first" industrialization but as an orderly system of graduated deviations from that industrialization' (op. cit.: 44).

Skip stages possible

In spite of these differences, which clearly point to the weakness in the modernization paradigm, we could possibly consider Gerschenkron's contribution as 'normal science' (showing dangerous anomalies) within the modernization paradigm. The problem of underdevelopment, as we know it today, had no place in this paradigm. There was only an original stage of backwardness, on which should follow a process which released the forces of modernization. These forces were seen as inherent in all societies and in so far as there was need for a theory of underdevelopment, its function would be to analyse 'barriers to modernization' and 'resistance to change'. Gerschenkron realistically pointed out the multilinearity of this process, but even in his analysis one finds the normative position that countries, after having achieved their takeoff, which Gerschenkron termed Big Spurt and which could be compared with Rosenstein-Rodan's Big Push, should return to the 'normal' track of development.

Political development and political order

Rostow's stages were basically derived from the distinction between 'tradition' and 'modernity' which is well known from classical sociology and the Weberian analysis of ideal models.

Maine described the transition between the two states in terms of *status* v. *contract*, Durkheim spoke of *mechanical* v. *organic* solidarity and Tönnies about *Gemeinschaft* v. *Gesellshaft*. There is no denying that more or less sophisticated versions of this paradigm were produced, but in its more simplistic form the modernization paradigm served as a development ideology, simply rationalizing cultural colonialism. Nowhere was this as clear as in the field of political science, where the tradition of studying change had been much weaker than in sociology on which it heavily relied. On the whole this genre was dominated by North American scholars (Almond and Coleman 1960, Almond and Verba 1963, Riggs 1964, Almond and Powell 1965, Pye 1966, Weiner 1966). Their framework was partially derived from the pattern variables and Parsonian systems thinking mentioned above.

Gabriel A. Almond, chairman of the American Social Science Research Council *Committee on Comparative Politics*, was the intellectual leader of the new movement and the major work showing the way was *The Politics of the Developing Areas* (1960), edited by Almond and Coleman. This work dealt with comparative politics rather than political development. Its weakness was the artificiality in delineating a specific political subsystem from society as a whole. Most seriously, the state was abstracted away.

The 'developmental approach' came later in a work that Almond wrote together with C. B. Powell (Almond and Powell 1965). Political development was here seen as an aspect of the wider process of modernization, marked by three criteria: structural differentiation, subsystem autonomy, and cultural secularization.

Other authors stressed different criteria but most of them saw political development as a complex concept (i.e. it included several dimensions) and as a component of modernization. For this reason they were trapped in the same teleological illusion as the sociological modernization theory. The content of political development or political modernization was implicitly identified with the institutional differences between the Western democracies and various traditional political systems. Political scientists who went out to study the 'developing areas' in the 1960s thus borrowed freely from their more experienced relatives: economics (politics of development) and sociology (political modernization). Naturally these field experiences had a sobering effect on theory.

A study which may be said to start from the assumptions of the modernization paradigm, while at the same time partly contradicting it, was David Apter's *The politics of modernization* (1965). He made a basic distinction between a 'reconciliation system' and a 'mobilization system'. The former was seen as corresponding to a modern pluralistic political system, whereas the latter constituted a transition phase between traditional and modern. Having worked in Africa, Apter realized that this transition implied immense social tensions and therefore a more or less dictatorial political organization was needed. This can be compared to Gerschenkron's 'processes of substitution' during the Big Spurt. The mobilization system and the Big Spurt should consequently be seen as temporary deviations from the normal, evolutionary path, and the idea of 'back to normalcy' is implicit in both theories. This idea reveals the impact of the modernization paradigm but at the same time the recognition of abnormalcy indicates a divergence from the paradigm.

Subsequently the interest among political scientists focused more on crises in development and nation-building. Five crises of development were identified: legitimacy, identity, participation, penetration and distribution (Binder et al 1971). It is interesting to compare this state-centred perception of crisis, facing single political systems in their development, with the present discussion of crisis, which is seen in a global context.[29] This is a good illustration of the transformation of development theory from the 1960s.

Of the liberal optimism implied in the early political modernization approach not very much remains today. This reflects of course the general decline of optimism in development studies. Almond himself later referred to 'the missionary and Peace Corps mood' prevailing during the 1950s and early 1960s (Almond 1970: 21).

In a critical review Donal Cruise O'Brien has pointed out that an increasingly more tense international and domestic situation has aroused the academic interest in political order rather than political change (Cruise O'Brien 1979). If 'modernization revisionism' (Huntington 1971; Randall and Theobald 1985) was more appreciative of indigenous social structure and culture (Rudolph and Rudolph 1967) it retained the evolutionary optimism of the classical paradigm. Traditionalism was seen as playing a positive role in the process of modernization in providing consensual elements not available in advanced societies. The stronger emphasis on political order in the late 1960s was on the other hand basically pessimistic and reflected a growing number of development catastrophes, military coups and political conflicts during the first development decade, particularly in Africa. The concept 'political development' went out of use (Riggs 1981). Social change seemed to be dangerous, not inherently positive, and political order was a necessity to prevent retrogression (Huntington 1968). Modernization appeared as a guided process rather than as natural history. From this conservative position there was but a small step to the view that the military rather than the bourgeoisie was the modernization agent, just to mention the most obvious example of the change in emphasis from political development to political order.

A more recent trend in political analysis is the political economy approach, also known as the public policy approach (Higgot 1983). Here the cross-disciplinary influences run from economics rather than sociology, and neoclassical rather than structural economics. It thus fits well with the counter-revolution in development economics, with its emphasis on choice and decision-making. Gone are concepts such as 'system', 'function', 'structure' and 'political culture', not mentioning 'political development'. Instead the ambition is to raise the scientific level, rigour, and precision of political analysis to match the more advanced economic analysis, even if the research territory by implication must be more limited.

> The grand theory of the 1960s can be characterized as explanation covering a wide number of cases with low explanatory power; the new policy oriented approaches of the 1970s are an attempt to provide a form of analysis which covers few cases with more effective explanatory power.
>
> (Higgot, op. cit: 27)

Undoubtedly the new approach can illuminate certain political realities (Bates 1981) but it neglects the larger historical-structural context as well as the cultural specificity of different contexts. The focus is on individual actors maximizing their interests and the analysis is based on universalist assumptions that ultimately tend to coincide with Western values.

2.3.2 Transitions to socialist modernity

Whereas both non-Marxist and Marxist theorists have identified 'a natural history' towards capitalism, only the latter apply the same logic to the establishment of socialism. In Marx's conception the road linking the original happy state of communism with the future, not only happy but also wealthy, state of communism, led through stages (cf. Rostow), and the mechanisms taking a society from one stage to another were inherent in the internal contradictions of each stage, representing a mode of production.[30]

The one historical transition which makes sense within this theoretical framework is the transition from feudalism to capitalism in European history, although there has been a great debate even among Marxists about the correct interpretation (Hilton 1978). When it comes to the transition from capitalism to socialism the confusion is complete. The problem was never theorized by Marx himself and there is little agreement about what is to be meant by 'socialism' (or for that matter 'communism' to which socialism is supposed to lead). Secondly no such transition, roughly corresponding to the Marxist scheme, has ever occurred. The major communist revolution, i.e. the Russian, and the type of society which emerged after that event has, as it were, filled our rather empty theoretical boxes with a content: the Soviet model. As theoretical definitions became state authorized the whole tradition froze into dogma. As we shall see in Chapter 6 it is, in view of the recent reform debate, legitimate to ask the question whether socialism might not be a transition to capitalism! However that may be, the goal of modernization is perhaps the only thing that escaped reinterpretation and reconsideration.

What is of crucial importance in the 'transition approach' is the emphasis on the internal contradictions within each mode of production paving the way for new relations of production. This in principle excludes the possibility of copying systems of production. The limited relevance of the Soviet model in the context of Third World development is thus clearly revealed in the variety of socialist development experience during the last decades.[31]

The first step towards this socialist pluralism was of course the break between China and the Soviet Union in the late 1950s when a 'Maoist model' was formed. Much could be said about this now discarded model, but the only point I want to make in this context is that Maoism was another 'quick transition' ideology, in this case related to 'the great leap' from socialism to communism. Furthermore, the Maoist view of the Soviet Union was that it had failed to achieve even socialism, and instead was retrogressing to a form of state capitalism (Gurley 1976; Riskin 1987). In Marxist terms such differences in opinion are serious indeed.

Soviet hostility to Maoism does not mean that Soviet development theorists fail to realize the need for modifications in the socialist development strategy. As a step in this direction the concept of the 'noncapitalist path' was launched in the early 1960s. The Soviet model itself did not include revolution, since it emerged in a post-revolutionary situation. However, it was long taken for granted that revolution was a sine qua non for the model to be implemented. The idea of a 'noncapitalist path', however, explicitly stated that socialist development was possible without going through the whole stage of capitalism and revolution if – and this was the main requisite – the Soviet Union gave its support to the country embarking upon the noncapitalist path, which in brief meant socialism without revolution.

The development strategy implied in this theory or ideology (it has served both functions) basically conformed to the Soviet model. Of course collectivization of agriculture was not possible with this reformist strategy, but the emphasis on industrialization, the state-sector and land reform was retained. Many dubious cases have appeared on (and disappeared from) the Soviet list of countries following the noncapitalist – or socialist-oriented – path.[32]

Both Soviet scholars and Western Marxists seem to have

greater doubts about the possibilities of creating 'socialism in one country' in the Third World today. Some Marxists focus wholly on the transition to capitalism, whereas others – refusing to exclude the possibility of a more voluntaristic option – define the transition as a fundamentally open process:

> . . . socialist solutions to underdevelopment do not exist as formulas that can be picked off the shelf, or magical balms that can be brewed up on the basis of the correct reading of old texts. Rather, if there is to be a new and successful social formation that can properly be called socialist, it will result from a long and complex process of transition that will, in turn, be deeply marked by the historical specificity of the societies undertaking the experiments. (Fagen et al. 1986: 10)

Not surprisingly, the idea of 'direct transition' has exercised a great attraction on socialists in different historical situations around the world. Even Marx had to face the issue of transcending the principle of orderly historical development in his correspondence with Russian populists, one of the great debates in Russian development thinking. The Latin American dependency tradition and the Soviet concept of noncapitalist development are variations on this theme: How much voluntarism is compatible with scientific socialism? Voluntarism invariably provokes defenders of the doctrine of orderly development: an Engels, a Plekhanov, and more recently a Bill Warren. We shall return to each of these debates in the contexts where they belong.

2.3.3 Criticism and rethinking

Growth and modernization theories were of course not completely dominant during the 1950s and 1960s, nor did they escape criticism. The modernization paradigm has been subject to strong criticism from social scientists in the Third World, particularly in Latin America. In an influential essay from 1966 the Mexican sociologist Rodolfo Stavenhagen attacked what he called the 'Seven Erroneous Theses on Latin America' (Stavenhagen 1966). The first of these stated that the Latin American countries were dualistic, consisting of one traditional agrarian, and one modern, urbanized society. The former was associated with feudalism and the latter with capitalism, which implied that feudalism was an obstacle to development that must be replaced by a progressive capitalism. Both societies were, however, in reality the result of the same process. In Stavenhagen's criticism of this and other 'erroneous theses', assuming for example automatic spread effects of industrialism, and a progressive role of the bourgeoisie, he and other Latin Americans provided a fresh departure in development theory: the dependency approach.

Brazil was, perhaps, the country in which the development optimism (desarrollismo) of the 1950s had found its most uninhibited expressions. Everyone, right across the political spectrum, thought that Brazil was in the takeoff stage, and that behind this achievement were the growing number of entrepreneurs. In São Paulo the Center of Industrial Sociology was established, where the various industries in the São Paulo area were studied in a Schumpeterian perspective. The results were unexpected: the Brazilian businessmen did not turn out to be the backbone of the growing Latin American bourgeoisie; instead they were found to be totally devoid of initiative and energy, totally dependent on the government and foreign capital (Cardoso 1967: 94–114). Doubts arose about the progressiveness of Latin American bourgeoisie, instead Gunder Frank coined the phrase 'lumpenbourgeoisie' (Frank, 1972).

Fernando Henrique Cardoso, who made the studies of the entrepreneurs in São Paulo, later wrote a general critique of the social sciences, particularly the 'theory of modernization' within sociology (Cardoso and Faletto 1969: 8–10).

Moving from Brazil to Chile we find the same disappointment at the way in which the established social sciences explained Latin American reality as well as their inability to provide guidelines for development policies. Osvaldo Sunkel claimed that the prevailing analysis of development was based on conventional theories which saw the mature capitalist economy as the goal of all development efforts. The underdeveloped countries were analysed in terms of a previous and imperfect stage on the way to this goal. Sunkel wanted this idealized vision to be replaced by a more historical method, the result of which would be a better understanding of the real nature of the underdeveloped nations' structures. The approach meant that the characteristics of underdevelopment should be viewed as normal results of the functioning of a specific system. The conventional theory considered these characteristics to be deviations from the ideal pattern which, like children's diseases, would disappear with growth and modernization. But, according to Sunkel, this would continue for as long as development policies attacked the symptoms of underdevelopment rather than the basic structural elements that had created underdevelopment (Sunkel 1969, 1973).

In this context it is difficult to ignore André Gunder Frank's brief but influential study The sociology of development and the underdevelopment of sociology from 1969, where he primarily attacked the Research Center on Economic Development and Cultural Change and its journal Economic Development and Cultural Change. Frank's Latin American experiences had obviously lead him to question the paradigm of this journal. His critique was directly aimed at the group around EDCC, e.g. Manning Nash, Bert

Hoselitz, Marion Levy, Everett Hagen and David McClelland. Frank went out to show that the modernization perspective, as developed by the above-mentioned theorists, was (a) empirically untenable, (b) theoretically insufficient and (c) practically incapable of stimulating a process of development in the Third World.

The thrust of the modernization approach was according to Frank to compare an underdeveloped country with a developed one by means of various indicators; the differences thus revealed were then established as the substance of development. This approach was manifested in two ways: by pattern variables and by stages of growth. As shown above the tradition of pattern variables goes back as far as the classical sociology, and was applied to the problems of underdevelopment by Bert Hoselitz.

Frank argued that many developed countries show strong particularistic tendencies, that ascribed status is widespread and that the structure of roles is not as functionally specific as the official ideology might have it. Similarly, traits of universalism, achievement and specificity might be found in the social structures of the underdeveloped countries.

Growth stages were a further development of the pattern variable analysis in the sense that the two idealized poles were united through a series of stages. As the reader may recall, Rostow mentioned five such stages: (a) the traditional stage, (b) the pre-takeoff stage, (c) the takeoff, (d) the road to maturity and (e) the mass consumption society. It is difficult to find these stages in reality. According to Frank underdevelopment was not an original stage, but rather a created condition; to exemplify he pointed to the British deindustrialization of India, the destructive effects of the slave trade on African societies and the destruction of the Indian civilizations in Central and South America. The greatest problem in Rostow's analysis was, however, the fact that not all of the countries, which according to him were ready for takeoff, could manage the final jump.

The theoretical shortcoming of Rostow's analysis was primarily the fact that it was based on 'comparative statics' rather than being dynamic, and that the overall perspective was lost. In terms of development policy the approach was gravely compromised because of Rostow's political affiliation:

> As to the efficacy of the policy recommended by Rostow, it speaks for itself: no country, once underdeveloped, ever managed to develop by Rostow's stages. Is that why Rostow is now trying to help the people of Vietnam, the Congo, the Dominican Republic, and other underdeveloped countries to overcome the empirical, theoretical, and policy shortcomings of his manifestly noncommunist intellectual aid to economic development and cultural change by bombs, napalm, chemical and biological weapons, and military occupation?
> (Frank 1969)

Frank's article was strongly polemical, as is illustrated by the quotation above; it was typical of the intellectual climate at the universities of the metropoles during the late 1960s when development studies took shape. In fact, development studies can hardly be understood if treated in isolation from this intellectual and political context.

After the critical foundation had been laid by the Latin Americans the modernization paradigm came into disrepute even in the West and there have been quite a number of analyses of its rise and fall (Smith 1973, Roxborough 1979). This critique does not only concern modernization theory as such, but the whole tradition of evolutionism and functionalism of which it forms part. Smith divides his criticism into four different lines of attack. Methodologically neoevolutionism is based on comparative statics which neglects both the sources and the route of change. From a logical point of view there is, for example, the mistake of equating serialism with causal explanations of transitions. Empirically it is easy to point out that any effort to classify societies using indicators of tradition and modernity soon breaks down. From a moral point of view, finally, the most clear-cut objection is the unabashed ethnocentrism implied in the modernization approach which, as we have seen, was one reason for the Latin American critique paving the way for the dependency approach.

On the winding road towards a theory of social change there are many mistakes, both magnificent and more trivial. Modernization theory certainly belongs to the first category. If one can speak of a paradigm in development theory the modernization perspective provided one. It had a long tradition in Western social thought. It was, mainly thanks to the doctrine of endogenism, logically coherent. It dominated several social sciences in the 1950s and 1960s. It had a great appeal to a wider public due to the paternalistic attitude toward non-European cultures. This in turn created the rationale of development aid, as well as the forms it took. It is probably correct to say that the general outlook of modernization theory still constitutes the popular image of developing countries.

What happened to the modernization paradigm within academia? In accordance with what was said about the tendency of social science paradigms to accumulate rather than fade away, we should be prepared for a come-back. Books critical of, but also rethinking modernization are still being published. In 1977 the outstanding mouthpiece of the modernization paradigm, Economic Development and Cultural Change, issued a 460 page supplement with essays in honour of Bert Hoselitz (Nash 1977). The naivety of the early formulations of the paradigm is here eliminated and much of the criticism referred to above is acknowledged. Rather than abandoning the approach altogether one has, as Manning Nash puts it, started to 'rethink modernization' (ibid: 28).

The rethinking can, it seems, take two opposite directions. The first possibility is to universalize the concept even further to distinguish it from the concrete manifestations of Western modernization. Wilbert Moore, for instance (in his contribution to the work mentioned above), reconceptualizes modernization as rationalization:

> If modernization is defined as the process of rationalization of social behaviour and social organization one avoids the vagueness of 'joining the modern world' and the ethnocentric connotations of 'becoming just like us'. (Moore 1977: 33)

More recently Nash defines modernization as 'the growth in capacity to apply tested knowledge to all branches of production' and modernity as 'the social, cultural, and psychological framework that facilitates the application of science to the processes of production' (Nash 1984: 6). He claims that this definition leaves open the actual, concrete modes of institutionalizing the modern values and does not stipulate the appropriate form of the social structure. This is essentially modernization without Westernization.

The other route is to conceive modernization as a historical process of Westernization. This view is not teleological; instead the process of modernization is seen as a consequence of the dominance of Western civilization during a certain historical period.

The crucial question then is the length of this historical period. There are reasons to believe that the end of it is close or at least that a decisive turning point is being reached. If modernization is understood as a project carried out by modern elites in the three worlds we must also consider what forces of demodernization there are, how strong they are, and what possibilities there are for these forces to combine. The secret behind the success of modernization is that the modernizers in the capitalist West, the socialist East and underdeveloped South are united, not directly (through a common purpose) but indirectly (through a common perspective). The factors indicating that a turning point is at hand are three:

- dysfunctional tendencies emanating from the modernization process itself;
- growing strength of antimodern forces;
- increased possibilities of antimodern coalitions.

Of these the third factor is the most problematic. We may distinguish three types of antimodern forces (Friberg and Hettne 1985):

- the traditionalists who are dependent on nonmodern structures and who tend to see the undermining of these structures as a threat to their way of life (peasants) or even as a threat to their physical survival (aboriginals);
- the marginals who are kept out or pushed out from the modern sector (the unemployed and the unemployable);
- the post-materialists who search for and defend values that have been eroded by the modern project (the green movements).

That the importance of these three categories in different respects is on the increase is not so difficult to see. The postmaterialist green movement is a new and growing phenomenon (although related to the counterpoint tradition) and so are the marginals in the industrialized world. Since full employment is being abandoned as a political goal this group can be considered a more or less permanent 'B-team' in the industrialized world. Although probably a shrinking force, the traditionalists have been around for a long time (all modern revolutions so far started as peasant revolts). Furthermore, a striking comeback is staged in some areas of Islam, and recently the plight of the aboriginals has become an international issue, due to the transnational organization of the 'Fourth World'.

The extent to which the antimodern forces can forge an alliance is a more complicated question but to take the case of the aboriginals again, the World Council for Aboriginal Peoples would have been hard to envisage a couple of decades back. The new social movements led by postmaterialist elites are basically transnational but so far mainly confined to the West. The marginalized groups belonging to neither of the two sectors (modern or traditional) are the most unpredictable and may support any strategy which promises to end their misery. It must be emphasized that a certain historical outcome does not necessitate a conscious and formal alliance. It is sufficient if the actors are departing from similar assumptions and sharing similar experiences.

I shall, however, abstain from making any predictions here. The point I want to make is simply that abandoning the evolutionistic and deterministic modernization paradigm opens up several options for future development, which it is the task of development theory (freed from teleology) to explore further. Such an analysis must also consider the world context of both modernization and 'demodernization'.

Chapter 3

The voice of the third world

> Development thinking within the social sciences is largely a product of the West. It is as such an outsider's view of our development, specially by outsiders from countries who colonized us. SUSANTHA GOONATILAKE

The statement quoted above, unfortunately still relevant, constitutes a great challenge to development theory. The major debate on the issue of cultural imperialism and the need for intellectual self-reliance in the Third World took place in the 1970s. At present it has receded into the background, partly because of a resurgence of conventional development thinking in the 1980s, but also because the voices from the Third World did become both louder and more articulate, thus to some extent correcting the imbalance.

This chapter first deals with the critique of the state of the art, as Third World intellectuals saw it in the 1970s, and describes some corrective measures on the institutional level. Secondly an account is given of a major theoretical school of thought with most of its roots in the Third World: the dependency school. Thirdly the discussion of indigenization is widened from the dependency debate in Latin America to cover broader social science issues throughout the Third World.

3.1 Academic imperialism and intellectual dependence

The process of indigenization, in development theory as well as in the social sciences as a whole, is fundamentally a movement of liberation from the colonial legacy and the imperialist world system, naturally reflected in the international pattern of communication. The first remedy has been to establish counterinstitutions, 'liberated zones' for critical and independent intellectual work.

This also implies an increasing dissonance within international social science.

3.1.1 Colonialism and social science

The social sciences necessarily reflect social conditions. To the extent they were applied in 'the colonial situation' (Balandier 1951) they became an integral part of colonial administration. This is particularly clear in the case of social anthropology, which did in 'primitive cultures' what sociology was supposed to do in advanced societies. The colonial roots of anthropology were intensely debated in the 1970s (Asad 1973). There also existed something called 'colonial economics' catering to administrators in the colonial services. The links between colonial economics and development economics can be illustrated by the fact that one of the pioneers (W. A. Lewis) in the latter field was appointed Reader in the former subject at the London School of Economics in 1947. When, much later, sociology and political science began to be applied in the Third World, the context was different: the issue of administration had been replaced with the issue of development. This, in a way, further underlined the Eurocentric bias, since the path to development was the imitative one, or what Narindar Singh has termed 'the Apean way' (Singh 1988).

The connections between the colonial social sciences and development studies are complex. Notwithstanding the fact that many old prejudices were transferred and new ones were added, the context was of course radically different. Most of the pioneers in development studies actually shared a feeling of Third World 'nationalism' or 'Third Worldism', and there was an evident guilt complex involved. It is significant that part of the recent criticism against development studies is an attempt to eradicate this guilt complex and re-evaluate the impact of colonialism.[33]

In this context one could also interpret some other, earlier discussed, theoretical countermovements within development-oriented social sciences, such as the stress on political order and the 'new political economy' (in its criticism of the Third World state) as a 'recolonization' of development studies or as a return of Eurocentric paternalism. At least such an interpretation is unavoidable in the Third World.

The problem of 'academic imperialism' was intensively discussed in the 1970s as part of the general debate of dependence. The critics saw:

> . . . research teams . . . move into their country with already designed research projects trying to 'mine' for data and statistics, using locals for semi-skilled activities like interviewing, filling out forms and interpreting, but preserving

for themselves the basic research design, processing and publishing. The 'researched' country, having been stripped of its data, sees the results published in the journals or books of the industrial countries, adding prestige to the foreign professors and their institutions. (Streeten 1974: 9)

The social sciences as practised all over the world, whether in government administration or in more or less autonomous university institutions, reflected the global structure of imperialism and dependence. Many social scientists in the periphery therefore tended to draw the conclusion that Western theories must be rejected, either because they were 'Western' or because they were 'bourgeois'.

Statements to this effect were commonplace throughout the 1970s in most countries of the Third World, particularly in countries where Western intellectual penetration has had a long history. One of the meetings of the faculties and schools of economics of Latin America, for example, did resolve that:

The analysis of the problems of Latin American development requires its own theory, which [. . .] should arise basically from the systematic observation and analysis of Latin American problems. The theory of development formulated in highly industrialized countries does not adequately explain such problems, and consequently cannot serve as a basis for a strategy and a policy capable of dealing with them successfully. (Cockroft *et al.* 1972: 308)

The necessity of generating an indigenous and relevant social science was also emphasized at conferences on Asian social science held in the early 1970s. On one such occasion M. S. Gore, Chairman of the Indian Council of Social Science Research, made the following observation:

I will probably not be mistaken if I state that our reference groups and the 'significant others' to whom we address our work are mostly located in Europe and the United States. This is not unnatural and can be explained by the much longer histories of social science development in the Western countries. But this Westward orientation tends to give rise to an unreal situation in which we fail to recognize the merit and the contribution of our scholars until they have been honoured abroad and, what is worse, accept them uncritically once they have been noted in some Western reviews. (Atal 1974: 28–9)

This criticism also applied to the socialist 'Apean way'. During the Hundred Flowers campaign unique documents about the internal social science debate in China became available. Many economists were dissatisfied and in a 'manifesto' the following critical points were raised:

Our science of economics as it stands at present is still retarded at a rather infantile stage; apart from transplanting in a doctrinaire manner Soviet Russia's textbooks and the like, we have nothing to show beyond description of the existing system.

Due to our treating the classical Marxist-Leninist works as mummified

> words, we are in the habit of putting the cap of Revisionism on whatever school of thought which happens to differ from the stand taken by the classical works. . . . This in turn blocks whatever possibilities there are for creative development of the science of economics.
>
> (MacFarquhar 1974: 177–20)

The problem of social science in Africa has been stated in much a similar way:

> Whether we like it or not social science is value-loaded and so long as Africa continues to rely on foreign social scientists it will continue to run the risk of unwittingly entrenching the very foreign culture and ideologies which it is seeking to supplant. (Temu 1975: 193)

The statements reproduced here are all examples of a general problem neatly summed up by Syed Hussein Alatas in the concept of the 'captive mind': the product of higher institutions of learning whose ways of thinking are dominated by Western thought in an imitative and uncritical manner (Alatas 1972). Thus, social science research tends to be a process of reproduction of Western values and, consequently, the Western model of development. This is only one way of stating that the social sciences are largely irrelevant in a Third World context. If we assume this as a fact, a major question emerges. Why is an irrelevant activity like social science research carried out in the first place?

Claude Ake, a Nigerian social scientist, has put the question bluntly:

> Why does Nigeria accept or tolerate a social science which is alienated from its environment and all but incapable of scientific development or of advancing significantly the struggle against underdevelopment?
>
> (Ake 1979: 17)

Ake's question refers to the Nigerian case but has more general validity. First, the educated higher strata of Nigerian society have internalized the values of Nigeria's colonizers. Second, most Nigerian social scientists have been trained in Western social sciences which implies they have a vested interest in the continuation of this type of training and research. Third, in so far as Western social science propagates the value of maintaining the existing social order, it is defended by the ruling class (ibid: 19).

The question of relevance can therefore not be discussed in isolation from the social context. A paradigmatic change in the social sciences certainly presupposes some kind of 'cognitive dissonance' but the gap between theory and reality is not a sufficient condition. Theoretical changes must be supported by changes in the power structure.

It is obvious that Western-style social science in general and Eurocentric development theory in particular have been, and still are, widely challenged in many academic centres in the Third World. The problem, which of course had been referred to by Third

World intellectuals long back, was most intensely discussed in the first part of the 1970s, and for natural reasons the discussion coincided with the debate on self-reliance, since indigenization of development thinking is an intellectual component of the general problem of self-reliant development.

Yogesh Atal has identified four stages of indigenization (Atal 1981):

- teaching in the national language and use of local materials;
- research areas carved out by domestic researchers;
- establishment of an academic infrastructure and national definition of research priorities;
- new and appropriate theoretical and methodological orientations making possible the creation of indigenous paradigms.

Obviously the problems increase towards the end of this chain.

3.1.2 The corrective movement

The universities in the Third World were established at different times and for different reasons in the various parts of the Third World. The earliest Latin American universities were founded in the sixteenth century mainly to protect the medieval Christian culture of the settlers in the new and 'barbarian' environment. This meant that the universities, as part of the conservative Creole establishment, were attacked by the liberation movements in the nineteenth century and had to be completely remodelled after independence.

In the case of Asia, universities were established during the later part of the nineteenth century, in Japan and China to provide what was known as 'Western learning', and in India to create an army of clerks and minor officials required for the cheap and smooth running of the machinery of British colonial administration.

If Latin American universities signified the beginnings of colonialism and the Asian universities its consolidation and reproduction, African universities heralded the eclipse of direct and overt Western dominance and its transformation into a neocolonial power structure. Unlike universities founded elsewhere by the colonial powers the African ones were established on the tacit assumption that direct rule was not to last for very long (Friberg *et al.* 1979: 17).

The function of universities in the Third World countries today is in a process of change from an instrument of intellectual reproduction to an instrument of development. Such a function is different not only from the colonial situation, but also from the traditional role of universities in Western countries. However, the change must not be exaggerated. Furthermore, if and when the role of the university as an instrument of development is accepted, a more universal question arises: What kind of development? Here an innovative and critical function of the university is suggested but at present this is very far from being the case. In fact, the notion that the university should play a role in development may work against its critical function, since the development function often seems to imply a subordination to government. On this issue a report noted some years back:

> On government–university relations, the Latin Americans express the most concern that universities are threatened by those who would make them an arm of government. The Africans view the problem as less critical, more concerned perhaps by the staggering need for development. The Asians take a position someplace between these two views.
>
> (Thompson and Fogel 1976: 8)

On the whole the academic freedom in the Third World has not improved. Rather the contrary. Besides, universities still largely function as institutions for mass production of professionals such as bureaucrats, lawyers and engineers. These difficulties have led to various institutional responses at various levels. The most immediate step seems to have been network building in order to strengthen indigenous social science and facilitate the horizontal (or South–South) diffusion of ideas.

Another reason for the creation of new institutions and networks is that policy-oriented research by necessity is interdisciplinary, whereas university teaching departments tend to be rather unconducive to interdisciplinary research. Therefore social science institutes of a new kind (interdisciplinary and problem-oriented) were created particularly in the 1960s and 1970s. In fact the more original contributions to development theory tend to come from these independent research institutes, rather than university departments. At the same time such institutes depend heavily on external financing. Therefore their originality, to some extent at least, is a function of the encouragement of new approaches by the funding agencies. This illustrates the complex relationship between intellectual penetration and intellectual emancipation.

The same can be said about the emergence of associations in development research which also often are supported by external agencies such as UNESCO or social science research councils and research funds in the rich world. In Latin America a regional council for social science, Consejo Latino-Americano de Ciencias Sociales (CLACSO), with headquarters in Buenos Aires was formed as early as 1967. In Asia a regional organization specifically concerned with development research within social science, the Association of Development Research and Training Institutes of Asia and the Pacific (ADIPA) was formed in 1971 at a conference in

Bangkok. In 1973 the Council for Development of Economic and Social Research in Africa (CODESRIA) was established with headquarters in Dakar.

Interestingly, these three regional associations, all concerned with development studies, were formed without the existence of a European model. Not until 1975 (on a suggestion coming from the Third World) the European Association of Development Research and Training Institutes (EADI) was founded, at present with headquarters in Geneva. In the following year the four regional associations decided to set up the Interregional Coordinating Committee of Development Associations (ICCDA). In 1977 it was joined by a fifth regional association, Association of Arab Research Institutes and Centres for Economic and Social Development (AICARDES), with headquarters in Tunis.

There are also Third World based networks such as Third World Forum and the Association of the Third World Economists. The Third World Forum was established at a meeting of leading Third World social scientists in Santiago de Chile in 1973. In a statement the group declared:

> What is required is nothing short of an intellectual revolution. This intellectual revolution must be carried to every university, every institute of learning and every thinking forum in the Third World.

From this brief survey some observations can be made. First, the development of a social science infrastructure is a recent process and, as far as South–South cooperation is concerned, it dates from the 1970s. Secondly, most of the new initiatives tend to be rather critical of prevailing paradigms in social science and there is a strong urge to develop more relevant approaches. Thirdly, and somewhat paradoxically, these initiatives have received a lot of support from different funding agencies in the centre which gives them a rather ambiguous status.

The impact of this development upon international social science on the whole remains to be seen. My impression is that the economic world crisis with corresponding fiscal crises in most Third World countries have been very damaging to social science collaboration and social science work on the whole.[34] The promising trend started during the 1970s later lost momentum. One major breakthrough, however, must be dealt with in more detail.

3.2 The rise of dependencia

In the literature on development and underdevelopment published in the 1970s there was one dominant perspective: the dependency approach. It originated in the extensive Latin

American debate on the problems of underdevelopment, and became a significant contribution to modern social science. Not only did it contain a devastating criticism of the Eurocentric modernization paradigm, it also provided an alternative intellectual perspective, rooted in the Third World, and it also functioned as a catalyst in the subsequent development of development theory. This new perspective went beyond the problem of structural dependency; it implied a self-reliant approach to development thinking as such, an indigenization of development theory. However, the parentage was a rather mixed one.

3.2.1 Origins of the dependency school

The dependency school emerged from the convergence of two major intellectual trends: one with its background in the Marxist tradition, which in turn contained several theoretical orientations: classical Marxism, Marxism-Leninism, neo-Marxism; the other rooted in the Latin American structuralist discussion on development which ultimately formed the CEPAL tradition.

The concept of neo-Marxism reflects a transformation of Marxist thinking from the traditional approach, focusing on the concept of development and taking a basically Eurocentric view, to the more recent approach, focusing on the concept of under-development and expressing a Third World view. Aidan Foster-Carter (1974), who coined the term, pointed to the following differences:

- Marxism (as interpreted by Lenin) sees imperialism in a centre perspective; neo-Marxism, on the other hand, sees imperialism from the periphery's point of view.
- The Marxist analysis of classes is based on specifically European experiences and emphasizes the emancipatory mission of the industrial proletariat, while the neo-Marxists have a considerably more generous view of different groups' revolutionary potential, for instance the peasantry.
- Marxists retain a somewhat deterministic emphasis on the objective conditions. The neo-Marxists view the possibilities of starting a revolution with greater optimism and emphasize the role of the subjective factors.
- Marxism still shows traces of the nineteenth century development optimism and considers the concept of scarcity to be a bourgeois invention for the purpose of legitimizing economic inequality. Neo-Marxists integrate the growing ecological consciousness with its view of development.

Of course a lot of controversy has arisen as to the continuity or discontinuity of these two approaches. To a growing number of more orthodox Marxists the main body of neo-Marxist thinking was seen as more or less incompatible with the classic Marxist framework (Taylor 1974). In other words the neo-Marxist or dependency approach was rejected as non-Marxist.

This labelling game may not be so important. What is important in this context, however, is that Marxism, which in its classic form was biased by the Eurocentric perspective discussed in the previous chapter, has become increasingly relevant for understanding the social reality in underdeveloped countries. For this to happen, a number of more or less fundamental changes in Marxism were bound to take place. Marxism as a Eurocentric world view had to become indigenized in order to become universalized. The problem is how fundamental changes were needed, and thus where Marxism ends and neo-Marxism begins. It can be shown that also in the later tradition of classical Marxism the prospects for industrialization were in fact considered as increasingly slim, if not altogether ruled out as in the subsequent dependency position (Palma 1978).

In spite of certain statements (with doubtful theoretical status) on contemporary issues, it is clear that Marx and Engels thought of capitalist development as a process bound to take place in a similar way in one country after another. In what now is known as the Third World this process needed the assistance of colonialism – repulsive, but yet historically progressive. Lenin held on to this evolutionary perspective in his analysis of the development of capitalism in Russia, although he emphasized that the process was complicated and full of contradictions, due to the country's backwardness. He nevertheless rejected the populist view that capitalism was impossible and must be bypassed. This view of capitalist development in backward areas influenced some dependency theorists, notably Fernando Henrique Cardoso. In his later work on imperialism Lenin in fact further modified the classical view, as he did not foresee any capitalist development in the colonies as long as they remained colonies. Instead the colonial bonds prevented the bourgeoisie from fulfilling its potential historical mission of releasing the productive forces. This interpretation certainly differed from that of Marx. At Comintern meetings during the 1920s the authoritative Marxist view even went a bit further in complicating the issue by its discovery of a feudal-imperialist alliance, which even in a post-colonial situation would provide a major obstacle to (capitalist) development. Imperialism contradicted capitalism in the Third World, which again was far from the original Marxian view. The underlying evolutionism, however, was never completely abandoned.

This perspective is inherent in all traditional Marxist writings on imperialism and Third World problems as well as in the more recent 'neoclassical' revival (Warren 1980). The Marxist literature on development and underdevelopment that emerged after the Second World War shared this perspective but increasingly a rather different type of analysis emerged. Underdevelopment was now seen as a continuous process rather than as an original state (to be overcome by development), and capitalist penetration was singled out as the principal cause of underdevelopment. This was the thrust of the theories of the North American Monthly Review group, particularly Paul Baran (1957).

Baran's views of underdevelopment were offshoots from his general analysis of US capitalism as an irrational system, both with regard to its internal logic and its effects on the Third World. The focus of his interest was how the actual surplus was produced and used compared to the potential surplus of a rational economic organization (socialism). The colonized areas were deprived of their economic surplus through mechanisms of imperialist exploitation. This was the origin of underdevelopment. Baran thus completed the chain of revisions in Marxist theorizing about imperialism with regard to its historical progressivity, and could therefore be seen as the founder of neo-Marxism, and consequently to some extent the dependency school. It was left to Andre Gunder Frank to bring this neo-Marxist line of thought to its logical conclusion in his thesis on 'the development of underdevelopment' (Frank 1966).

Thus within a broad Marxist framework imperialism has been both hailed as a promotor of development (in its capitalist form) and accused of being the creator of underdevelopment. This contradiction within the Marxist tradition has for obvious reasons generated some confusion and, following from that, a tendency to abandon the concept of imperialism altogether. Outside the Marxist framework interpretations of 'imperialism' did proliferate further, and it has become a popular exercise in semantics to display the various concepts and their more or less obvious incompatibility. This is not particularly fruitful since each generation perhaps should be entitled to its own brand of imperialism.

A study by Giovanni Arrighi (1983), 'disturbed by a loss of common language,' takes a less relativistic approach and, building on the Hobson paradigm, tries to reconstruct a concept of imperialism which is free from the usual contradictions. Arrighi's intention was not to propose a new theory of imperialism, but to analyse the premises of current and classical theories. Rather than presenting a definition of imperialism, valid for all times and historical situations, he develops a theoretical space ('a geometrical metaphor') leading to four varieties of expansionary trends or 'imperialisms'. To fill Arrighi's theoretical space takes a lot of empirical explanation that cannot be undertaken here. Suffice it to underline Arrighi's conclusion that the classical theories on imperialism are totally irrelevant for the study of the capitalist world-economy since the Second World War.

So far we have dealt with the tradition of Marxist imperialism theory. A second important background to the dependency school was, as mentioned above, the indigenous Latin American discussion on underdevelopment, reflecting specific economic and intellectual experiences in various Latin American countries, particularly during the economic crisis of the 1930s. The depression dramatized the dimensions of Latin American dependence, it initiated more systematic economic research (e.g. by the central banks), and it necessitated a policy of import substitution, later systematized into a fully-fledged development strategy. In the 1950s this strategy was generalized into a Latin American Strategy by the Economic Commission for Latin America (ECLA or, in Spanish, CEPAL).

Before dealing with this strategy we must mention the two most prominent Latin American structuralists: Raúl Prebish and Celso Furtado. The late Raúl Prebisch was the most prominent critic of outward-oriented development and his ideas concerning industrialization through the strategy of import-substitution can be traced back to his writings on Argentina in the 1930s and 1940s which were reminiscent of the European tradition of economic nationalism (Love 1980). Continued reliance on the export of primary products would only consolidate the 'peripheral' position of the developing countries in the world economy. Just like List, Prebisch wanted to promote export. A necessary precondition, however, was a developed industrial base. This could only be created through a temporary seclusion, not necessarily on the national but possibly on a regional level. Prebisch lived through decades of development debates and of course he continuously modified his views. In his last speech, at the ECLA session in 1986, he made a plea for a combination of import substitution and export promotion (Singer 1986). One thing which Prebisch consistently stressed throughout his works was the world dimension in all economic processes, expressed in his simple but forceful Centre-Periphery model (Sunkel 1986).

Celso Furtado similarly took the Great Depression and its effects on the Brazilian economy as a point of departure for a new approach in economic analysis.[35] In particular Furtado was impressed by the need for historical understanding of the development process. His later contributions to Latin American structuralism were related to the problem of inflation. The emphasis on noneconomic factors, as well as the internal-external dynamics, led to the theory of dependence. The point to be stressed in this context is the link between Latin American structuralism and the theory of dependencia.

Latin American structuralism came to be known through the structuralist–monetarist debate on inflation in the 1950s and early 1960s. To the structuralists the phenomenon of inflation was not related to the money supply but to various inelasticities and institutional rigidities which only could be tackled through structural reforms. Thus the structuralist recipe against inflation became a development strategy popularized by CEPAL. The monetarists on the other hand had to wait for their turn.

Since there are many valuable accounts of CEPAL (Hirschman 1961, Cardoso 1977, Rodriguez 1980, Pollock 1978) a brief summary may suffice here. The CEPAL-doctrine may not appear as particularly radical in view of the more recent dependency debate. It is therefore in order to recall the very hostile climate in which CEPAL was born. When the United Nations during the post-war years committed itself to the problem of economic reconstruction with the establishment of ECE (Economic Commission of Europe) and ECAFE (Economic Commission for Asia and the Far East) the Latin American countries wanted to set up their own regional commission. The major opponent of the 'regionalization' of the United Nations was the United States, interpreting the Latin American demand as a declaration of independence. In spite of this resistance CEPAL was established in 1948 with headquarters in Santiago de Chile. The hostility of the United States increased when the theoretical and policy position of CEPAL ultimately was crystallized (Prebisch 1950). In the context of conventional development thinking in the 1950s the CEPAL-doctrine was conceived as revolutionary by some and as utopian by others (Pollock, op. cit: 61).

What then was this doctrine?

- Theoretically, it was an assault upon the conventional wisdom concerning the relationship between international trade and development, and contained the elaboration of an alternative framework: the centre-periphery model. According to this perspective only the 'central' nations did benefit from trade, whereas the 'peripheral' nations suffered. There were several reasons for this: trends in the terms of trade, political asymmetry, technological factors, etc.
- In terms of development strategy, the CEPAL doctrine emphasized industrialization by import substitution, planning and state interventionism in general, and subsequently regional integration was strongly emphasized.
- Ideologically the CEPAL doctrine constituted the most recent example within a long tradition of economic nationalism,

starting with F. List and the German reaction to the dominance of Britain as the workshop of the world.

The remedy was thus thought to be industrialization based on import substitution, by which the import of various consumption articles was replaced by domestic production. This implied protection during an initial phase and also a certain coordinative (and in fact increasingly interventionist) function performed by the state. As stressed above, such a strategy, modest as it may appear today, was highly unorthodox, almost revolutionary, in those days. Nevertheless, during the 1950s the CEPAL doctrine became accepted as the proper development strategy by many Latin American regimes. Implementation was a different matter.[36]

For a limited period import-substitution worked rather well but later experiences showed that the strategy was, if not wrong, at least inadequate.[37] Its inadequacy can be explained by two factors. First of all the industrial process necessitated inputs which had to be imported and therefore created another kind of dependence, technological and financial. Secondly the pattern of income distribution in Latin America confined the demand for manufactures to a relatively small elite, and as soon as it had been satisfied the growth process came to an end.

The CEPAL economists, working under institutional and political constraints, were reluctant to draw the obvious conclusions from this experience. The resulting state of cognitive dissonance provided incentives for elaborations on the dependency approach, which resulted in a variety of 'dependency schools', some of them continuations of the old CEPAL strategy, others more oriented towards Marxism of some kind.

Unlike the concept of imperialism which had been imported from abroad, dependencia was largely an indigenous Latin American creation. However, the concept of dependency also forms an important part of general theory and must be analysed in that context as well.[38]

3.2.2 Varieties of the dependency approach

In view of the complex intellectual origins of the notion of dependency, including Marxism (or rather Marxism-Leninism), neo-Marxism and Latin American structuralism, there are of course several conceptualizations to choose from.[39] They differ in style, emphasis, disciplinary orientation and ideological preferences, but they nevertheless share the basic idea about development and underdevelopment as interrelated processes. This perspective, which was a clear break with the modernization paradigm, is particularly clear in the following definition by T. Dos Santos:

> Dependence is a conditioning situation in which the economies of one group of countries are conditioned by the development and expansion of others. A relationship of interdependence between two or more economies or between such economies and the world trading system becomes a dependent relationship when some countries can expand only as a reflection of the expansion of the dominant countries, which may have positive or negative effects on their immediate development. (Dos Santos 1970: 231)

This way of formulating the dependency perspective certainly gives emphasis to economic processes and may also be interpreted as if dependency were a one-way causational chain. In fact, the 'dependentistas' put more emphasis on internal factors compared to the 'cepalistas'. Dos Santos later elaborated on this problem in order to avoid any misunderstandings about the respective role of external and internal dimensions. He made a distinction between conditioning and determining factors, in saying that the accumulation process of dependent countries is conditioned by the position they occupy in the international economy but determined by their own laws of internal development (Dos Santos 1977). The result is nevertheless a dependent capitalism, unable to break the chains with metropolitan centres and achieve its full development.

Cardoso compares this position with the Russian narodniks, who had asserted that capitalism in the late nineteenth century was impossible due to the limited internal market. Against this position Lenin made the point that capitalist development was a contradictory process, characterized by social tensions and destruction. This, however, did not make capitalism impossible in Russia. Nor, according to Cardoso, was capitalist development (although of a dependent kind) impossible in Latin America.

Dependency theory went beyond economics in attempting to provide a general explanation of underdevelopment, and the radical dependentistas in particular were often sociologists. They therefore conceived dependency as a sociopolitical phenomenon, which in turn made room for an even more complex view of the centre-periphery relationship. In a preface to the American edition of their classic work on dependency, Cardoso and Faletto say:

> We conceive the relationship between external and internal forces as forming a complex whole whose structural links are not based on mere external forms of exploitation and coercion, but are rooted in coincidences of interests between local dominant classes and international ones, and, on the other side, are challenged by local dominated groups and classes.
> (Cardoso and Faletto 1979: xvi)

For Osvaldo Sunkel, drawing on Latin American structuralism, development and underdevelopment were two interrelated processes of one single structure, transnational capitalism, which became increasingly integrated, whereas national economic systems, particularly in the periphery, underwent a process of disintegration. Thus the aim of development was greater national autonomy (Sunkel 1969).

Do such variations in the choice of formulations signify deep theoretical differences? In my opinion, characterizations and summaries of the dependencia approach have been oversimplified, partly because they were written with a polemical intent, and partly because they identified the theoretical positions of the dependency school by means of a one-dimensional analysis, i.e. by stressing one distinctive factor while neglecting others. Thus the dependency theorists have failed to recognize their own theoretical contributions in the criticism, which consequently has been of little constructive value.

> The most general and formal of Gunder Frank's works are received as though they were his best, the formal definition of dependency furnished by Theotonio Dos Santos is appended, the problematic of 'subimperialism', and 'marginality' is sometimes inserted, one or another of my works or of Sunkel is footnoted, and the result is a 'theory of dependency' – a strawman easy to destroy. (Cardoso 1976: 13)

Instead of creating a strawman it would perhaps make more sense to locate various theoretical positions within a framework containing several relevant theoretical dimensions such as:

- Holism v. particularism. On this dimension two types of theorist are contrasted: those working with global models, the dynamics of which are determined by the system in its entirety, and those who establish the overall perspective starting from its constituent parts. For instance Sunkel's model of transnational capitalism has a holistic ambition, whereas Cardoso's view of dependency as a method for concrete analyses of the periphery is more particularistic.
- External v. internal causal factors. The fact that the distinction is difficult to make does not concern us here. The question is, which factors are the more important? Of course no dependentista would admit neglect of internal factors, but nevertheless the whole approach had an externalist bias (as is illustrated by Frank's metropolis-satellite model). After all, this was the main objection to modernization theories. When Augustin Cueva (1976) on the other hand stated that 'it is the nature of our societies that determines the links with the capitalist world' he articulated a position which placed him outside the camp.
- Sociopolitical v. economic analysis. Some theorists worked exclusively with economic analysis; others emphasized social and political conditions. Although this could be explained by

disciplinary origin, the difference was nevertheless important to the mode of analysis. The CEPAL tradition was on the whole rather economistic, whereas many dependentistas, as noted above, came from other social sciences, e.g. sociology.

- Sectoral/regional contradictions v. class contradictions. Some authors accentuated the fact that a regional or sectoral polarization occurs in the total system, both at the international and at the national level; others based their analysis on the assumption that the fundamental conflict is to be found in class contradictions. In other words, the latter sought the dynamics in the class struggle. For example, in his model of global dualism, Sunkel stressed cleavages between marginal areas and the transnational core (incorporating parts of centre, parts of periphery). These cleavages cut through classes, thus working against class consciousness and class struggle. Cardoso on the other hand gave more emphasis to class, but his analysis was subtle, going far beyond the simple dichotomy of capital-labour and the official communist position. On the whole, however, there was very little class analysis in most of the dependency writings.
- Underdevelopment v. dependent development. A central argument of the dependency school was that a situation of dependence generates a process of underdevelopment. However, some assumed a more cautious attitude, claiming that stagnation tendencies are a cyclic problem, and that development of capitalism is fully compatible with a position of dependence.

  The strong dependency position was formulated by Frank in the phrase 'development of underdevelopment'. The more cautious position is Cardoso's notion of 'associated-dependent development'.
- Voluntarism v. determinism. The great majority of the dependency theorists did see their research as politically relevant. However, a distinction can be made between those who considered the political means constrained by the objective situation, and those who stressed the possibility of overcoming these limitations by direct, political action. The latter attitude is obviously linked to the idea that Latin America (and the Third World) was doomed to underdevelopment and that political activism was the only response. Typically this activism took the form of guerrilla struggle and the model was provided by the Cuban revolution. Consequently those who admitted the possibility of some development, albeit along capitalist lines, took a middle of the road position between the official communist and the extreme voluntarist points of view.

From this discussion it should be possible to construct an ideal-typical dependency position which, whatever relevance it might have in reality, does express a certain internal consistency. A typical dependency position would stress holism, external factors, sociopolitical analysis, regional contradictions, polarization between development and underdevelopment and the role of subjective factors in history. The contrary position on the various dimensions would fall outside the dependency school and in fact define a mode of production approach to which we return in the next chapter.

If we elaborate on such a 'cluster' and thus minimize theoretical diversity, the dependency theses concerning underdevelopment and development could be seen as the following:

- The most important obstacles to development were not lack of capital or entrepreneurial skills, but were to be found in the international division of labour. In short, they were external to the underdeveloped economy – not internal.
- The international division of labour was analysed in terms of relations between regions of which two kinds – centre and periphery – assumed particular importance, since a transfer of surplus took place from the latter to the former.
- Due to the fact that the periphery was deprived of its surplus, which the centre instead could utilize for development purposes, development in the centre somehow implied underdevelopment in the periphery. Thus development and underdevelopment could be described as two aspects of a single global process. All regions participating in this process were consequently considered as capitalist, although a distinction was made between central and peripheral capitalism.
- Since the periphery was doomed to underdevelopment because of its linkage to the centre it was considered necessary for a country to disassociate itself from the world market, to break the chains of surplus extraction, and to strive for national self-reliance. In order to make this possible a more or less revolutionary political transformation was necessary. Politics would take command. As soon as the external obstacles had been removed, development as a more or less automatic and endogenous process was taken for granted.

On this level of generalization dependencia can be seen as a paradigm. There will naturally emerge a basic agreement on a higher level of abstraction, while theorists as a typical feature of 'normal science' tend to differ when it comes to more concrete issues. As this general perspective was accepted, previous approaches in social science, such as growth models, pattern variables and political modernization, drastically lost relevance. Compared with the endogenism of the modernization paradigm, the dependency approach in its stress on external factors or the impact of the world context appears almost as an antithesis. However, with respect to the content of development the difference is slight.

Dependencia thus can be seen as a new point of departure rather than a new theory. This was, by the way, explicitly recognized by some of those Latin American theorists who originally developed this approach to underdevelopment, for example F. H. Cardoso, who was closer to mainstream Marxism. On the other hand Frank, Dos Santos and Marini at one stage made attempts at building a more formalized and autonomous theoretical tradition based on dependencia. Sunkel, Paz and Pinto continued in the more pragmatic CEPAL tradition. Thus, the dependency school always contained different lines of thought, while at the same time expressing a number of common ideas.

These ideas, which so clearly emerged from the empirical reality of Latin America, constituted the most formidable challenge that the Eurocentric concepts and theories on development had faced so far and were received with enthusiasm by intellectual circles throughout the Third World. They also had a strong impact on Western scholars working in the area (notably Gunder Frank) and began to conquer the Western academic community from the late 1960s onwards. In this way two conflicting paradigms in development theory arose (Foster-Carter 1976a).

Of course the applicability of the concept of paradigm to social science (held by Kuhn to be in a preparadigmatic stage) is somewhat problematic, but the Kuhnian notions of normal science, crisis and scientific revolution should undoubtedly be of relevance in this context. That the two schools of development theory ('growth and modernization' and 'dependence and underdevelopment') constituted competing and partly incompatible paradigms is shown by the obvious lack of communication, often taken to be one important criterion of paradigmatic conflict.

This communication gap can be studied in the editorial policy of various academic journals. The new perspective as well as the theoretical critique was reflected in the contributions to journals such as Journal of Contemporary Asia, Review of African Political Economy and, of course, Monthly Review. Gradually it got a fairly generous treatment in World Development, Journal of Development Studies and Journal of Peace Research. Somewhat surprisingly, one article by Dos Santos is found in American Economic Review, but this appears to be an exceptional case (Dos Santos 1970). Dependencia was of course never able to conquer the stronghold of the modernization paradigm: Economic Development and Cultural Change. However, some self-criticism among modernization theorists was, as we have seen, influenced by the dependency perspective.

3.2.3 Criticism and assessment

It is supposedly the fate of every paradigm to be questioned, attacked and, ultimately, replaced. Dependency certainly is no exception as became evident in the late 1970s when Colin Leys declared:

> It is becoming clear that underdevelopment and dependency theory is no longer serviceable and must be transcended. (Leys 1977)

The evidence for Ley's harsh judgement was (a) theoretical repetition and stagnation, (b) the existence of problems which the theory cannot solve, (c) an evident lack of practical impact. Among more specific objections were:

- The meaning of development is obscure.
- It is unclear whether it is the underdeveloped countries or the masses in these countries that suffer from exploitation.
- Concepts like centre and periphery are nothing but polemical inversions of the simplistic pairings of conventional development theory (traditional-modern, etc.).
- The theory tends to be economistic in the sense that social classes, the state, politics, ideology get very little attention.
- The ultimate causes of underdevelopment are not identified apart from the thesis that they originate in a centre.

This critique, summarizing a debate which in fact had gone on throughout the 1970s, proved to be quite influential, perhaps also because a personal conversion was implied. Leys himself had used the dependency approach in his earlier studies on Kenya (Leys 1975). Another reason may be its sweeping and general character. Obviously not all varieties of dependency are equally vulnerable to the objections, but as we noted above the nuances are seldom known outside the inner circle.

From a methodological point of view the concept of dependence was attacked for circular reasoning and for having limited explanatory value. In an effort to distinguish dependent from nondependent countries a number of problems arise (Lall 1975). In reality almost all countries – even those we normally do not conceive as underdeveloped – do import technology, are dependent on exports, have a tendency to emulate consumption patterns in other countries, contain marginalized groups and regions within their territory, and so on. Thus, the exercise of distinguishing 'dependence' from 'nondependence' soon breaks down as long as countries are the units of comparison.

This also implies that it would be difficult to rule out the possibility of industrialization for a certain category of countries based on this distinction. As became clear in the early 1970s a number of countries in the Third World were in fact industrializing at a high speed, contradicting a generally held view that their development was blocked. Of course it could always be argued that they, in spite of being industrialized, remained dependent on a higher level, but this tautological argument gradually lost credibility. Even those countries that today are fully industrialized went through this process in the context of the world market on which they were 'dependent', and with very differing preconditions in terms of indigenous technology and capital. It has convincingly been argued that the dependency theorists had an exaggerated picture of the self-centred nature of classical capitalist development.

> Underdevelopment theory cannot have it both ways. If the field of analysis is world economy, if the centre needs the periphery for modes of exploitation that off-set the tendency of the rate of profit to fall, if the circuit of capital in general is realized on the international plane, then there is no capitalist formation whose development can be regionally autonomous, self-generating or self-perpetuating. 'Development' cannot be conceptualized by its self-centred nature and lack of dependence, nor 'under-development' by its dependence and lack of autonomy. (Bernstein 1979: 92)

Circularity of reasoning thus stands out, retrospectively, as the most glaring methodological deficiency of the dependency tradition (Booth 1985).

The criticism of the more simplistic forms of dependencia on the theoretical level is found also within the dependentista camp. In several articles Cardoso expressed his dissatisfaction with 'the ceremonial consumption' of the theory, particularly in North America. He noted that the dependency paradigm has generated a new set of erroneous theses about Latin American development, albeit very different from the seven theses exposed by Rodolfo Stavenhagen in his classical essay on the modernization paradigm.

The development of capitalism at the periphery was, according to the strong dependency position, prevented by structural obstacles, as for example the limited internal market and lack of dynamic capital. In Cardoso's view it was wrong to elevate this idea to the level of law. The existence of contradictions is not an insurmountable obstacle to capitalist development, but rather a characteristic of such development. Conjunctural phenomena must therefore not be translated into permanent characteristics of peripheral capitalism.

Cardoso also questioned the impotence of the bourgeoisie in peripheral countries and its incapacity to fulfil its 'historic mission', a central theme in dependency theory. To speak of a 'lumpenbourgeoisie' as Frank did, was to go a bit too far. The idea that the Latin American continent faces a choice between socialism or fascism was also disputed by Cardoso. Although he did not deny that the dominant classes militarized their style of domination, this did not necessarily imply a fascist political organization. Instead one could now speak of a new social category, the 'state bourgeoisie' – 'a social stratum that politically controls the state productive apparatus, in spite of not having private ownership of the means of production.' This stratum could evidently encourage hope of expansionist statism, quite independent of the interests of the multinationals, and it would therefore be more accurate to speak of 'preimperialism' rather than 'subimperialism'.

This discussion reveals the complexity of the dependency approach as well as the elaboration and deepening of the discussion which took place in Latin America in spite of the difficult political situation in most countries. Thus, the intellectual milieu created in Santiago de Chile was destroyed after Allende, and efforts were made to re-establish a new centre in Mexico. Here the discussion for some time continued both on a theoretical level, for example between those who wanted to place dependency analyses within the Marxist tradition and those who preferred a more eclectic approach, and on the empirical level, where a growing number of analyses of the state, the local bourgeoisies, the working class, political movements, marginalization, urbanization, patterns of industrialization, activities of the TNCs, etc., were initiated, all using dependency as their frame of reference. On the theoretical level there emerged three schools of thought – estructuralistas, marxistas and dependentistas – involved in what appeared to be a very fruitful discussion, which, however, also formed part of the disintegration of the dependencia school.[40] It had in fact been transcended.

To summarize the critique is quite problematic. Since the group of theorists classified as 'dependentistas' did not express identical views there have on various levels been critical debates both within the group and involving one or the other of them in polemics with external critics, which in turn can be classified as 'rightists' or 'leftists'. Many critics belong to more or less distinct schools in development theory and their criticism is indistinguishable from the general tenets of these schools. A systematic account of the critique therefore implies more or less an overview of development theory, which is the subject of this whole book.

It is commonly held that the dependency school has been proved irrelevant and generally rejected. This is not quite true and it is therefore important to distinguish between various levels of the debate: the paradigmatic, the theoretical, the methodological and the empirical (Werker 1985). The paradigmatic debate implies a collision of general assumptions and world views, including diverging theoretical orientations. We shall return to the debate on this level repeatedly in following chapters. On the theoretical level we find contributions which, in contrast with the paradigmatic critique, accept the assumption of blocked development but attempt to solve some of the shortcomings of classical dependencia, for instance the world-system and mode of production approaches, which also will be discussed later on. To the theoretical debate we must obviously add the debate within the camp, to which some references were made above. We also gave examples of methodologically oriented discussions, for instance on the difficulty of distinguishing 'dependent' from supposedly 'nondependent' countries.

Finally, a comment on criticism at the empirical level. The experience of the NICs (Newly Industrialized Countries) has been used as the main case for the refutation of dependency theory. At the same time countries trying to delink and follow the strategy of self-reliance have been less than successful. The empirical evidence is, however, quite intricate when scrutinized closer.

The impact of the dependency school should not be underestimated because of the criticism. It can be found in four areas:

- The decline of the modernization paradigm.
- The stimulation of dependency analysis in other areas of the Third World.
- The emergence of new development strategies.
- The catalysing effect on development theory.

The modernization paradigm was weakened by its inability to explain real developments in the Third World during the 1960s, as dependency theorists such as Stavenhagen, Cardoso and Sunkel were quick to point out. Stavenhagen questioned the assumption that the Latin American societies were dualistic. Cardoso noted that the idea of tradition v. modernity was derived from European sociology. Sunkel stressed that the idealized and mechanical vision of development ought to be replaced by a more historical method. Frank's *The sociology of underdevelopment and the underdevelopment of sociology*, summarizing some of this critique, was certainly one of the top list publications among the New Left in the late 1960s, and formed part of a general climate of intellectual change.

The dependency perspective also became popular among social scientists in other parts of the Third World, particularly among a younger academic generation searching for a critical platform. One could see the dependency debate in the Third World as a declaration of intellectual independence, paving the way for a process of indigenization of development thinking, to be further discussed below.

The dependency perspective has also influenced the discussion on development strategies both on a national level and on an international one (e.g. the discussion of collective self-reliance in the context of NIEO).

On the national level regimes influenced by the dependency perspective were Chile under Allende, Jamaica under Manley, and Tanzania under Nyerere. Significantly only one of these regimes survived (but modified its development strategy substantially), which indicates that self-reliance is a difficult option in the context of the present world order.[41] Of course it may be disputed whether dependency theory was of much importance to the regimes referred to. For example Pedro Vuscovic, Allende's minister for industries, explains that the dependency school, in contrast to CEPAL, did not formulate any concrete economic programme.[42] The lack of a theory of development was surely a great weakness in dependency theory. So much stress was put on the external obstacles to development that the problem of how to initiate a development process, once these obstacles were removed, was neglected. In fact one gets the impression that the development perspective implied in dependency theory was rather close to the endogenous growth and modernization paradigm. On the other hand the difficulties involved in self-reliant development were played down before the full extent of integration of the world-economy was revealed by the crisis of the 1970s.

The demands for a New International Economic Order (NIEO) articulated in the mid 1970s were partly related to the breakthrough of the dependency paradigm. Possibly it could be seen as the outcome of a marriage between a nationalist-bourgeois interpretation of dependencia, on the one hand, and the political influence suddenly achieved by the oil-producing countries on the other. The unprecedented success of OPEC was possibly a factor behind the vociferous demands for fundamental changes in the international economic order during the Sixth Special Session of the General Assembly in 1974. This is shown by the fact that the Algerian president Houari Boumedienne not only took the initiative to this conference but also played a leading role in the proceedings leading up to the famous Declaration on the establishment of a new international economic order. This document indicated a shift of emphasis in the work of the United Nations in the 1970s: Development was the priority item on the international agenda (Sauvant and Hasenpflug 1977: 4). This is not to say that the demands incorporated in NIEO were new, nor that much has been done to implement them. Most of them had been stated repeatedly since 1964 by UNCTAD under the leadership of Raúl Prebisch. What was new in the case of NIEO was the firm and well coordinated political behaviour of the Third World delegates which took the rich world by surprise.

The dependency paradigm stimulated the debate on NIEO and provided the critics of the old international order with many good arguments and an appropriate language. Possibly the CEPAL element in the dependency school was most important here. Some of the weaknesses and biases of the dependency approach also crept into the NIEO debate, e.g. economic nationalism, the stress on external factors and lack of concrete development strategy. It is therefore natural that the radical dependency theorists considered this as a case of conservative reabsorption. To Samir Amin the NIEO was 'a consistent and logical program for getting out of the crisis, that reflects the interests and views of the bourgeoisies of the South'.[43]

Turning to the theoretical impact of the dependency school its crucial role was to focus upon the particular conditions affecting the development process in the Third World and the many unforeseen contradictions that characterized this process. Its corresponding weakness was the exogenist bias, being an antithesis to the endogenist bias in the modernization paradigm, as well as the classical Marxist approach. Accordingly the responses to the disintegration of the dependency school have been attempts towards more global approaches, incorporating the complex relationships between both central and peripheral development.

According to A. G. Frank (1977: 357) dependence has now completed the cycle of its natural life, and the reason is the crisis of the 1970s. New theoretical approaches are called for:

> . . . the usefulness of structuralist, dependence, and new dependence theories of underdevelopment as guides to policy seems to have been undermined by the world crisis of the 1970s. The Achilles' heel of these conceptions of dependence has always been in the implicit, and sometimes explicit, notion of some sort of independent alternative for the Third World. This theoretical alternative never existed, in fact – certainly not on the noncapitalist path and now apparently not even through so-called socialist revolutions. The new crisis of real world development now renders such partial development and parochial dependence theories and policy solutions invalid and inapplicable. (Frank 1981: 27)

Thus the dependency school is no more. Its demise has left an awkward theoretical vacuum since its critics have generally been less successful in pointing out new theoretical directions. However, the debate contains a number of old and new approaches in which the catalysing role of the dependency school is more or less evident. One obvious example is the call for 'indigenization' in the 1970s.

3.3 The indigenization of development thinking

Dependency shifted the centre of the development debate towards the Third World – but not far enough. The radical solution to the problem of academic imperialism was to do away with Western concepts altogether and to build 'national schools'. This was by many conceived of as a threat to social science as such, since the result would be a fragmentation, not only between different disciplines of social science, which is a problem by itself, but also between particularistic and parochial national traditions. However, in a longer perspective attempts at indigenization of development thinking may be seen, and this is the view taken here, as a necessary phase in the growth of a more comprehensive and relevant concept of development. Such a concept would not necessarily preclude Western concepts, but implies a more realistic view of Western social science as reflecting a specific geographical and historical context. What follows is a discussion of selected efforts, mainly from the 1970s, at formulating more indigenous approaches to the problem of development in Latin America, Asia and Africa. This particular theme in fact deserves a book of its own, for which this brief survey is a poor substitute.

3.3.1 Latin America in search of otherness

In Latin America there is for historical reasons a strong cultural similarity with the West (the Latin subculture) and a high degree of academic institutionalization, particularly in countries such as Mexico, Argentina, Brazil and Chile. Western values have not been questioned to the same extent as in Asia and Africa.[44] What has been questioned is rather the capacity of mainstream social science to correctly describe the peripheral reality and to generate feasible strategies for change. This critical attitude is very similar to the one taken by the emerging nations of the nineteenth century, namely Germany and the United States. A case in point is the dependency tradition which to a certain extent functioned as a nationalist ideology.

The fact that Western development thinking on the whole seems to be integrated with Latin American thought does not mean that efforts to search for a specific historical identity are lacking. These, however, seem to be confined to writers and philosophers (pensadores). The famous Mexican poet Octavio Paz, for example, refers to the existence of a counterculture which he, for lack of a better word, calls 'the other Mexico'. Paz presupposes 'the existence in each civilization of certain complexes, presuppositions, and mental structures that are generally unconscious and that stubbornly resist the erosions of history and its changes' (Paz 1972: 75). In the case of Mexico this duality has become rather extreme, mainly due to the closeness of the US, a fact that has set the pattern for modernization in Mexico.

> The developed half of Mexico imposes its model on the other without noticing that the model fails to correspond to our true historical, psychic, and cultural reality and is instead a mere copy (and a degraded copy) of the North American archetype. Again: we have not been able to create viable models of development, models that correspond to what we are. Up to now development has been the opposite of what the word means: to open out that which is rolled up, to unfold, to grow freely and harmoniously. Indeed development has been a strait-jacket. It is a false liberation. (ibid: 73)

Where should one look for the other Mexico or, for that matter, the other Latin America? In preconquest civilizations? Many would reject these traditions as violent, repressive and inhuman, although it may be true what Paz says that 'it is a mistake to study the totality of Meso American civilization from the Nahua point of view (or, worse, from that of its Aztec version) because that totality is older, richer and far more diverse' (ibid: 88).

Another Mexican intellectual, José Vasconcelos, has defined the alternative in a more synthetic and universalistic way. He hoped for the emergence in America of a 'cosmic race', a 'new cultural being' that combined Indian, African, and European elements (Jorrin and Martz 1970: 216). Like Vasconcelos, Leopoldo Zea equates 'society' with the North American social system, whereas 'community' is given Latin or Hispanic connotations.[45] Thus the reference is regional rather than national, as far as the philosophy of otherness is concerned. There is also a strong assimilationist bias in these conceptions.

Innovations in the social sciences include the concept of *marginalidad* (marginality), developed by the Mexican social scientist Pablo Gonzáles Casanova (Gonzales Casanova 1965). Marginality is a rural phenomenon, implying extreme poverty and certain indigenous (in this case Indian) cultural traits which are fairly persistent. The fact that the various manifestations of marginality are correlated indicates an integrated phenomenon, which does not seem to be much affected by conventional development. The existence of a marginal population amounting to 10 million persons in Mexico raises some doubts about the validity of main-stream development theory with its insistence on the spread effects of economic growth. The case of Mexico is particularly relevant in view of the revolution with the subsequent structural modifications, and the continuous process of economic growth which has taken place in that country.

In the case of Mexico indigenismo played an important

political role both in the struggle for independence and in the Mexican revolution. The latter was a rather multifaceted process but it is obvious that Zapata's struggle in the state of Morelos was a struggle for the preservation of traditional village community (ejido). On the whole, however, these aspirations were frustrated and the main historical function of the revolution was instead to eliminate the obstacles to industrialism and urbanism (Ross 1975). This meant that the problem of marginalidad was a result of mainstream development, which in turn called for 'another' development or, in an ethnic context, ethnodesarrollo (Bonfil 1982; Stavenhagen 1986). The concept of 'ethnodevelopment' is discussed elsewhere in this book.

Since Latin America has had its fair share of military coups it is not surprising that there are significant contributions in this field, which may prove to be of wider relevance in social science. Of special importance in development studies is the debate around the theory of bureaucratic authoritarianism (BA), coined by O'Donnel (1973). This theory relates the new militarism which emerged in the 1960s and 1970s with a crisis in development, characterized by repression (the decline of populism) and outward orientation (the exhaustion of easy import substitution industrialization), leading to a 'deepening' of the productive structure. The framework for this theoretical debate (Collier 1979) was dependencia and the target was the modernization paradigm.

One method of bridging the cultural cleavage between the Eurocentric theories of the intelligentsia and the praxis of the nonelites is action research, as developed by Orlando Fals Borda in Colombia (Fals Borda 1970). Briefly it contains the following methodological steps:

- Analyse the class structure in the region in order to identify the groups which play a key role.
- Take from such key groups the themes to be studied with priority, in accordance with the conscience and action level of the groups themselves.
- Seek the historical roots of the contradictions that dynamize the class struggle in the region.
- Restore to those key groups the result of the research with the purpose of obtaining a greater clarity and efficiency in their action. (Quoted from Rudqvist 1986: 122)

Thus, action research has the double purpose of increasing the efficiency of political practice and to enrich social science. As a method it has a universal relevance. Along with dependencia it may be regarded as a distinctly Latin American contribution to social science, although there have been parallels elsewhere.

Marxism in Latin America, on the other hand, is a clear case of intellectual transplantation with little indigenous production, until the formation of the dependency perspective in the 1960s. The most important exception to the rule is the Peruvian José Carlos Mariátegui whose Siete Ensayos de Interpretacion de la Realidad Peruana (Seven Essays on the Peruvian Reality) written in 1927 remains 'the single most important attempt to understand a national, Latin American problem in a Marxist perspective' (Aguilar 1968: 12).

Mariátegui claimed that the Spanish conquest meant a retardation, and that the subsequent development of capitalism was perverted, partly by the foreign economic influence and partly by the domestic alliance between the bourgeoisie and the aristocracy. To Mariátegui the 'proletariat' was the Indian population, and the new Peru should be built on the latter's collectivist traditions.[46]

However, we know very little about the actual or potential role of the preconquest tradition in Latin American thought. In his introduction to Latin American thought H. E. Davies mentions that particular issue among the unresolved problems in this field of research:

> Do the religious, social, and esthetic concepts derived from the Amer-Indian and Afro-American heritage have useful meaning in the present-day ethnic scene? Do they have validity in relation to present problems, or are they a limiting, restricting inheritance, standing in the way of the self-realization of Latin American peoples? (Davies 1972: 234)

Before we leave the Latin American cultural area it is necessary to consider the discussion on development and underdevelopment in the Caribbean, which is remarkably little known in spite of its high originality and quality. Being so close to Latin America one would expect dependency theories to flourish, particularly if one considers the obvious relationship between a country's dependence and its size. The problem of size was, as will be further discussed in Chapter 5, a key issue in the Caribbean economic discussion (Demas 1965). In few areas of the world has the problem of economic integration been so thoroughly analysed. Another contribution to the theory of development and underdevelopment is the important discussion on the legacy of the plantation system (Beckford 1972).[47]

As far as the problem of dependency is concerned the opinions differ as to whether the Caribbean school of dependency is an autonomous development (Girvan 1973) or simply a reflection of the larger Latin American debate (Cumper 1974). It is not possible to prove any of these positions since both external influences and internal factors, which by themselves were sufficient for a theory of dependency to develop, did play a part. Among the latter may be mentioned:

- A history of extreme colonial subordination resulting in the so-called plantation economies, referred to above.
- A fairly developed academic infrastructure and a particularly advanced level of economics.
- Political conditions that were conducive for the ideology of dependency to develop, particularly after the Manley regime came to power in Jamaica in 1972.

It is possible, however, to trace the new approach further back to the formation of the New World Group in 1962 in Georgetown, Guyana. Its journal, the New World Quarterly (NWQ), later moved to Kingston, Jamaica. Among radical economists associated with NWQ can be mentioned George Beckford, Lloyd Best, Havelock Brewster, Alister McIntyre, Kari Polanyi-Levitt, Clive Thomas, and, of course, the two editors Norman Girvan and Owen Jefferson. Their writings on dependency and underdevelopment certainly are comparable to those of the Latin Americans.[48]

In comparison the non-European cultural elements seem to be stronger and so is the emphasis on indigenous forms of development (Nettleford 1978). Beckford considers phenomena such as Rastafari and Reggae as 'embryonic advances towards an indigenous social living'.[49] For Walter Rodney dependency was primarily white supremacy, the remedy being black power.[50] Thus, in the Caribbean, the non-European culture is more visible and has a more direct impact on the social science debate.

3.3.2 The sociology of civilizations: India and China

Since the discussion on dependency in Latin America should be seen as the major example of intellectual emancipation, a relevant question is whether there has been a similar debate in Asia. In fact there was a strikingly similar discussion in India as far back as the end of the nineteenth century. I am referring to the famous 'drain theory' of Dadabhai Naoroji. The thesis, presented in an early draft already in 1867 (Naoroji 1962), asserted that Britain secured a yearly tribute of enormous proportions from India. This unjust transfer of capital robbed India of her development potential in terms of infrastructure, education, etc. Thus a causation was established between the drain and the lack of development.

> Considering that Britain has appropriated thousands of millions of India's wealth for building up and maintaining her British Indian Empire, and for directly drawing vast wealth to herself; that she is continuing to drain about £30,000,000 of India's wealth every year unceasingly in a variety of ways; and that she has thereby reduced the bulk of the Indian population to extreme poverty,

> destitution and degradation; it is therefore her bounded duty in common justice and humanity to pay from her own exchequer the costs of all famines and diseases caused by such impoverishment. (Naoroji, op. cit., p. 622)

Naoroji's theoretical argument was obviously related to his political activity as an Indian nationalist – who lived most of his life in London – and his political objective was subsequently achieved. 'Drain acquired an almost hypnotic effect as a symbol of the injustice of British rule in all its manifestations. It gave an image to the multiplicity of Indian grievances' (Chandra 1975: 119). This could be compared to the role of the more recent dependency perspective in pointing out the effects of neocolonialism and preparing the ground for NIEO. In fact Indian economists have stressed the similarity between the drain theory of Naoroji and the so-called Prebisch–Singer thesis (Minocha 1970: 37). But the parallels do not end here. Compare the statement of Naoroji, quoted above, with the following thesis of Paul Baran:

> Thus the British administration of India systematically destroyed all the fibres and foundations of Indian society. Its land and taxation policy ruined India's village economy and substituted for it the parasitic landowner and moneylender. Its commercial policy destroyed the Indian artisan and created the infamous slums of the Indian cities filled with millions of starving and diseased paupers. Its economic policy broke down whatever beginnings there were of an indigenous industrial development and promoted the proliferations of speculators, petty businessmen, agents, and sharks of all descriptions eking out a sterile and precarious livelihood in the meshes of a decaying society. (Baran 1957: 149)

The neo-Marxist view that countries in the Third World, rather than following the development path of the advanced countries, were actively underdeveloped by the latter, not only in the colonial days but still today, can be derived from Baran's work. Since his major example was India, one is tempted to think that the drain theory of Naoroji could have been a source of inspiration. Baran did not refer to Naoroji directly but to other contemporary observers such as R. Palme Dutt and William Digby who made a similar point. Thus, to formulate a somewhat daring hypothesis, the old Indian debate on the drain might have influenced the Latin American dependency debate through Paul Baran.

However, there are differences between the drain theory and dependency theory to be noted. Naoroji's criticism concerned the 'un-British' colonial rule. Never did he suggest that Indian underdevelopment was a consequence of the 'free play of market forces'. On the contrary he pleaded for 'liberalization'.

Dependency theory as such was never very popular in India, which probably has something to do with the *déjàvu* effect and, of course, the size of the country.[51] Many Indian intellectuals tend to feel that India's economic problems are something more than external exploitation. According to Rajni Kothari:

> The dependencia theory is very relevant but it becomes an alibi for lack of self-development. You can always put the blame on the door of the exploiters but the exploitation that takes place in your own society is not questioned. However, unless the change takes place in the centre of the world the peripheries will continue to suffer at least culturally and intellectually, if not economically.[52]

The limited attraction of dependency theory is also illustrated by this comment in Economic and Political Weekly (23 April 1977: 666):

> There could be no greater slur inflicted on our capabilities: we are nincompoops, we are unable to ensure a local supply of exploiters, the process of exploitation has to be initiated elsewhere . . . This itself is neo-colonialism of a sort.

It is not easy to tell whether this reaction against overemphasis on external factors is 'indigenous' or whether it simply reflected the state of the international discussion. In those parts of the Third World where the dependency theory did establish itself, this reassessment is going on. However, in India, where Marxism was rooted early, one is inclined to think that the Marxist critique on the whole had an independent base.[53] This is amply demonstrated in the vigorous debate on the development of capitalism in India's agriculture in Economic and Political Weekly (Bombay) and Social Scientist (Trivandrum).

There was in the early 1970s a particularly vivid discussion in Contributions to Indian Sociology on the need for a special sociology of India,[54] as expressed by J. P. Singh Uberoi:

> Until we can concentrate on decolonization, learn to nationalize our problems and take our poverty seriously, we shall continue to be both colonial and unoriginal. A national school, avowed and conscious, can perhaps add relevance, meaning and potency to our science, continued assent to the international system cannot. (Uberoi 1968: 123)

Two decades have passed since that statement was made. What happened to the brave programme? As a tentative answer I would suggest that the swarajist viewpoint did gain strength in the early 1970s but that the discussion remained in a stage of confrontation – a confrontation that no longer is in the forefront of the more nuanced intellectual debate of today (Oommen and Mukherji 1986).

As concepts which specifically dealt with an Indian situation and which account for the complex cultural background one could mention R. Kothari's The congress system and M. N. Srinivas' Sanscritization. The first referred to the peculiar way of elite-formation and nation-building in India where the original elite groups kept themselves in power by selective recruitment and gradually extended control (Kothari 1970). The second concept simultaneously tried to explain the process of caste mobility and the intricate dynamic relationship between dominant Hindu culture and the parochial peripheral cultures of India (Srinivas 1968).

It is obvious that the dethronement of 'modernity' in development studies stimulated a more positive interest in what was hidden behind 'tradition'. As pointed out by S. C. Dube, 'tradition is a vast reservoir' (Atal and Pieris 1976: 85). The elements of tradition can be used to support the maintenance of status quo, but it can also offer sustenance to a radical reconstruction of society, as so often was stressed by Gandhians.

Thus one obvious reference point for the indigenization of Indian social science is Gandhi, whose struggle for political emancipation was, basically, a struggle for the restoration of Indian civilization. This is widely (but sometimes ritually) recognized among social scientists in India, particularly on the occasion of Gandhian anniversaries.[55] On many points Gandhi, in spite of his Hindu idiom, appears as strikingly modern. His approach may be described as action-oriented (the oppressive environment was his laboratory), normative (his viewpoint was that of the poorest of the poor) and global (the ultimate goal was a non-violent world order). Similar principles may be found in contemporary problem-oriented social science research.

Gandhi's views on development are, however, too big a subject to go into here.[56] One problem affecting the relevance of his ideas for development theory and social sciences in general has been the fundamentalist approach of many of his followers. A second problem is the inconsistency between a Gandhian development path and the Indian power structure. Since India became independent, planners have largely avoided experimentation with Gandhian development principles in spite of the high reverence for the Mahatma. In 1974 and 1975 the Indian political scene was dominated by the JP movement, a resurgence of political (as distinct from ritual) Gandhism. After a short spell of authoritarianism under Indira Gandhi the Janata government, supported by the Gandhian leader Jayaprakash Narayan, came to power in 1977 on a more or less Gandhian programme. Only a year later it was clear that the new government had failed and India returned to normalcy. In consequence the Gandhian alternative had suffered, possibly beyond repair (Hettne 1976). This also meant a setback for those social scientists who were more inclined towards indigenous approaches.

Perhaps an even harder blow to the idea of self-reliance is implied in the similar fate of Maoism in China. If Gandhi has been at least a symbol for the indigenization of the social sciences in India, Mao Zedong played a more direct role in China. In the latter country the Leitmotif in this intellectual process has been the contradiction between universalist scientism (associated with Soviet Marxism) and concrete activism (associated with Maoist philosophy). As part of Mao's goal of eliminating the 'three great differences', one of which concerned manual versus intellectual work, the academic nature of the social sciences was de-emphasized to the extent of being extinguished. Mao, however, always considered social studies of utmost importance for political work:

> No really good leadership can result from the absence of a real, specific knowledge of the actual condition of the classes in Chinese society. The only way to know the situation is to make an investigation of society, to investigate the life and condition of each social class. (Quoted from Wong 1975: 463)

Mao emphasized this 'action research' approach to the extent of identifying social science education with actual work on farms and factories. Thus, the demarcation line between scientific literature and party propaganda became blurred. The ritual, politically motivated insistence on Maoism as an indigenous source of knowledge reached its high-water mark during the cultural revolution.

There were two types of social investigations. The first one, which was therapeutic and educational in intent, was based upon the hsia fang ('sending down') system. Cadres, officials and scholars who had become 'divorced from reality' were sent to communes and factories in order to become re-educated. In this way the gap between mental and physical labour and between the cities and the countryside was supposed to be narrowed. The other type was more task-specific and usually performed by a research team. The purpose here was to get precise information on the various local conditions (Wong 1979). The prototype for this very problem-oriented research is of course Mao's Report of an investigation into the peasant movement in Hunan (1926).

No theoretical studies were allowed. The research should be pragmatic and policy-oriented. Obviously, as many Western commentators were quick to point out, this meant that the borderline between knowledge and propaganda got blurred, and that social studies often were perverted. However, this is only one aspect which may be countered with the functional importance of realistic insights into the social process at the grass roots level. The typical Chinese social investigation was a complex phenomenon, serving a number of important purposes in Chinese nation-building. It provides a fascinating example of aborted indigenization of social science. Today the social science agenda has, of course, changed completely.

If Latin American social science has made very important contributions to the problem of dependence, while (in relative terms) neglecting internal structure, Asian social scientists have largely been more interested in the internal possibilities and constraints of development, rooted in old civilizations.

3.3.3 The battle for decolonization in Africa

Africa is an example of more recent intellectual penetration and a comparatively weak academic infrastructure (Mkandawire 1988). The processes of intellectual penetration and intellectual emancipation have consequently been going on simultaneously although in different social contexts. The struggle for self-reliance in the academic field is still highly relevant (Mandaza 1988).

From the point of view of indigenization, a golden age of 'social science' in Africa was during the struggle for independence and in the immediate post-colonial era (Atta-Mills 1979). During this period original political ideas, many of them with a social science content, emerged. Take for example the various intellectual contributions from Kwame Nkrumah, Sekou Touré, Julius Nyerere, Leopold Senghor, Jomo Kenyatta, Patrice Lumumba, Frantz Fanon and Amilcar Cabral. What they produced was a form of non-institutionalized social science, comparable both to Latin America's pensadores and Maoist sociology. Certainly it was primarily politically oriented, dealing with the situation of colonialism, African identity, and alternative strategies for liberation. But who would assert that the colonial social science with its pragmatic concern with kinship structures, patterns of migration, traditional leadership, etc. was less politically motivated? Or, for that matter, the 'modern' development of social science with its obsession with modernization, planning and growth, legitimating the establishment of a bureaucratic elite and the strengthening of state power.

The first phase of indigenization did not survive the emergence of modern social science, and the second phase, which is more directly concerned with academic social science, has just begun. What are the similarities and differences between the two phases?

The earlier phase of African social and political thought, related to the Western challenge, was of course marked by the concrete experience of imperialism. Thomas Hodgkin has pointed out the difference in emphasis between Western (Marxist-Leninist) and African theories of imperialism. For the former the main issue was what made the capitalist world imperialistic, whereas the latter were more concerned with what imperialism did to Africa (Hodgkin 1972). This theorizing implied a reinterpretation of the civilization/barbarism theme of the Western apologists of imperialism, a reinterpretation which occasionally took the form of populist views about the undesirability of Western industrial civilization in general. A prominent example of this theme was the concept of African socialism, which, even if it should be looked upon primarily as a political and ideological concept, nevertheless did have theoretical significance (Friedland and Rosberg 1964). As an ideology it encompassed many diverse viewpoints (Klinghoffer 1969: 16):

- The Afro-Marxists emphasized Marxist-Leninist ideas of economic development and political structure. Major examples were Kwame Nkrumah and Sekou Touré during the first half of the 1960s.
- The moderate socialists, including Kenyatta of Kenya and Kaunda of Zambia, favoured a state-controlled 'socialist' economy but were at the same time anxious to attract foreign investment capital.
- The social democrats were closely connected with European socialism and frequently pro-Western in outlook, for example Leopold Senghor of Senegal and Tom Mboya of Kenya.
- The agrarian-socialists (populists) were associated with Nyerere's Ujamaa-philosophy. Rather than looking for foreign models of socialism Nyerere looked for it in traditional African society.[57]

Thus as the examples above show, African socialism covered a wide ideological spectrum from more or less pure Marxism-Leninism to populist ideas rather similar to the Russian narodniks or Gandhi in India, as well as nationalist ideology (Fanon, Cabral). It is the latter, more indigenous thinking, which is of most relevance in this context.

One example is the Senegalese economist Mamadou Dia, who tried to work out an alternative economic system, at the same time African and democratic-socialist, and preserving fundamental values of the traditional society (Dia 1960). The other varieties of African socialism referred to above were on closer examination rather conventional development ideologies inspired by the Western (including the Soviet) model, but dressed in African clothes to make them more acceptable to the people.[58]

Later in the 1970s trends toward intellectual emancipation, not only in the broader cultural sense but more specifically in the social sciences, became manifest. This could possibly be seen as inspired by the world-wide debate on dependence, which also had an African component in the 'Dar es Salaam School'. The criticism of status quo elsewhere typically took the form of Africanization, in terms of personnel rather than in terms of concepts and theories. This may be exemplified by developments in the Institute of Development

Studies (Nairobi) which originally was established as a 'neocolonial' institution, but later was gradually Africanized. This process did not primarily involve criticism of prevalent schools and traditions of research. As far as the process of Africanization did concern the content of research, the changes were limited to a stronger emphasis on pragmatic research objectives, which were instrumental from the point of view of the administration, rather than alternative perspectives. The dissatisfaction with the Western bias in university courses and research programs is widespread but, in general, teachers find it very difficult to find good substitutes which are more relevant in the African context.[59]

The analyses of dependency in Africa often came from expatriate researchers from Europe or the Caribbean.[60] A major exception was the Egyptian economist Samir Amin who in his more empirical phase produced many important studies of North and West Africa in a world context. In Africa Amin stood out as a dependency theorist in his own right. Although his indebtedness to early ECLA analysis was acknowledged, Amin has also emphasized discussions within the Egyptian communist party in the 1950s as influential.[61] Together with Gunder Frank and Immanuel Wallerstein he later became associated with world-system theory (Amin 1974, 1976, 1977) and thus more concerned with universalization than indigenization. He claims that the heterogeneity of peripheral economies disguises their underlying unity.

Also in Africa, however, social scientists are beginning to realize that the process of indigenization must go beyond both dependency theory and Marxism:

> The irony of dependency theory, in cultural terms, is that it is not necessarily in defence of national cultural autonomy. It is an argument for socialism rather than capitalism, an argument that anticipates a new social formation based on a new value system. The new social system and the value system which supports it will tend to converge with the value system of the East and diverge away from the value system of the West and the original value system of the changing African society. (Uchendu 1980: 93)

The current phase of indigenization takes as its point of departure the development crisis in Africa and a criticism of the conventional development paradigm. It is true that critical radicalism often has moved towards some kind of Marxism rather than reviving purely indigenous patterns of thought, but this may not necessarily imply an incompatibility between Marxism and more or less indigenous approaches. A case in point is the Ghanaian economist Tetteh A. Kofi who tried to develop the Abibirim strategy of development (Kofi 1974, 1975a, 1975b).[62]

Kofi was inspired by the Caribbean dependency discussion (both Kofi and the Caribbeans argued against W. A. Lewis's strategy of 'industrialization by invitation'), and emphasized the need for a new breed of economists who are capable of analysing not only the 'modern' but also the 'traditional' sector in order to formulate a strategy that is less disruptive or painful to the traditional society.

For a number of other African states, self-reliance is no longer a matter of deliberate choice, but is becoming a sheer necessity. As was discussed in Chapter 1 the post-colonial state is in a crisis and it is becoming more and more difficult to sustain the modern infrastructure. The traditional sector is pushing the modern sector back, not as part of any conscious strategy of self-reliance (such a strategy would have made the process less painful) but as a more or less spontaneous development, as was observed by F. W. Lukey during an acute phase of the crisis in Ghana:

> The quick solution has been tried and it failed miserably. The longer path – of a steady switch out of a city of traders and intellectuals, backed by illiterate farmers, using primitive techniques, into a modern participative state, where people engage in producing the things they need and things to trade, and apply themselves to increase efficiency of doing so, thus creating the wealth they desire – lies open. (Lukey 1978: 14)

Among those who fled the cities were social scientists. The economic crisis almost broke the back of African Social Science. The process of indigenization was in the first place brief and inconclusive, the honeymoon between the universities and the state soon ended, and when the IMF and the World Bank as part of the adjustment drive asked for reduction in social expenditures, social science was found to be expendable.

To conclude this theme it is obvious that the brave programme of indigenization is far from being fulfilled and that the whole issue is far more complicated than was realized in the 1970s.

If indigenization refers to the process in which transplanted ideas and institutions are more or less radically modified by the receivers to suit their own specific situation, it is not always clear to what extent this implies intellectual emancipation or new forms of dominance. It is for instance a fact that some funding agencies actually encourage intellectual emancipation. Often the receiving party does not represent a national culture on which indigenization can be based. In India for example modernization along Western lines would be replaced by Sanscritization, which according to non-Hindu groups implies a form of intellectual Brahmin penetration. In Africa, where the nation state in many countries has yet to become a social reality and where people identify with sub-national communities, it is not so obvious what an indigenous national development path would look like. In many Latin American countries, on the other hand, the process of national integration has gone further. Therefore the nationalistic function of the dependencia perspective and the strategy of self-reliance is more evident.[63]

However, even in Latin America the indigenization process is complicated by the existence of a Latin-Indian cleavage. The indigenous Indian culture (indigenismo) has inspired writers and artists, but so far it has had little impact on the social sciences. Furthermore, the idea of a synthetic Latin American culture presupposes the assimilation of the Indian subculture which in the current period of Indian awakening is unconceivable. The debate on indigenization in the 1970s was undoubtedly a promising start, but before the takeoff the debate somehow was nipped in the bud. One reason was probably that the project of indigenization was based on the assumed existence of more or less homogeneous national cultures. In many cases this proved to be an erroneous assumption. The nation state has been fundamentally challenged in different ways. The major trend during the 1980s has not been indigenization but globalization.

Chapter 4

The globalization of development theory

> We are increasingly confronted, whether we like it or not, with more and more problems which affect mankind as a whole, so that solutions to these problems are inevitably internationalized.
> THE BRANDT COMMISSION

The preceding chapters have dealt with two kinds of bias in development theory, endogenism and exogenism, which relate to each other as thesis to antithesis. If carried to their extremes, both approaches are equally misleading. The obvious remedy is to transcend the dichotomy and find the synthesis. In the real world there are no countries which are completely autonomous and self-reliant, and no countries whose development can be understood merely as a reflection of what goes on beyond their national borders. All countries are dependent on each other, and on the system of which they form part, but there are different types of dependence, both in kind and in degree. This fact is often described as interdependence, a concept lending itself to diverging interpretations:

- To some the notion is a refined dependency concept, suggesting a more complicated structure of the world economy than the simple centre–periphery dichotomy. Elements of this complexity are for example the rivalry and tensions within the centre (Western Europe being more 'dependent' than the US, Eastern Europe more so than the Soviet Union), the industrialization of the periphery and the deindustrialization of the centre (as exemplified by South Korea and Great Britain), and the emergence of regional powers (Brazil, India, Nigeria).
- To others the notion of interdependence suggests a common predicament for the peoples of the world ('we are all in the same boat'). This interpretation conveniently disregards the fact that the passengers of the boat (if we may continue the marine metaphor) do not travel in the same class, nor do they have the same access to the lifeboats.

Thus interdependence is an ambiguous conceptual innovation of relatively limited theoretical usefulness. Nevertheless it represents a challenge for current theories and may therefore serve as a point of departure for exploring various approaches to the analysis of development in a global perspective. The chapter concludes with a discussion of the changing role of national development strategies in the context of world development, increasing interdependence and decreasing room to manoeuvre for the state.

4.1 From dependence to interdependence

Interdependence is not only subject to contrasting interpretations but also contains distinct dimensions, for instance physical, political and economic. With physical I refer to the 'Spaceship Earth' discussion centred on the finite nature of global resources and ecology; with political the 'world order' debate; and with economic various attempts to come to grips with the changing international economy, no longer comprehensible in terms of trade relations only. We shall also discuss the experiences of the NICs (Newly Industrialized Countries) whose spectacular development is bound up with what goes on 'out there' in the world economy.

4.1.1 The emergence of planetary consciousness

One obvious dimension of interdependence is the physical-cum-biological and ecological notion of wholeness and finiteness expressed in concepts such as 'Spaceship Earth' and 'global commons'. During the 1970s in particular, several manifestations of a 'planetary' consciousness could be noted. The UN conference on the environment (1972) emphasized that ecological systems have little to do with national borders. Mankind is ultimately dependent on the biosphere, and therefore no country can upset the ecological balance without affecting other countries. There is, one began to realize, 'only one earth' (Ward and Dubos 1972).

Global modelling and physical systems theorizing thus became fashionable, and after decades of 'realist' conventional wisdom in command, explicitly normative 'world order models' were also considered a legitimate academic concern (Mendlovitz 1975). The world models most in focus in the 1970s were The Limits to Growth, Strategy for Survival, The Alternative World Model (Bariloche Model) and The UN World Model. They represented a new type of studies in which a set of computerized variables were analysed from the point of view of various theoretical assumptions in their relations to each other and the 'system' they defined. One decade later, however, they are mostly forgotten.

At the level of world order, the demand for a New International Economic Order (NIEO) and, at the turn of the decade, the proposals of the Brandt Commission Report were in focus. The common perspective underlying these documents, despite differing origins (the former from the South, the latter from the North), was one of impending crisis and ultimate breakdown of the world system, unless drastic reforms were undertaken. The strategy of global reformism implied in the NIEO proposals and the 1980 Brandt Commission Report (followed by a second report in 1983) thus also presupposed the 'one world–one system' approach. The key word in both Brandt reports was interdependence rather than dependence and, in contrast with the zero sum assumptions in the NIEO, the mutuality of interests was emphasized.

In this tradition, which goes far back in international relations theory, interdependence implies both a theory: an interdependent world promotes peace and development, and a strategy: interdependence should be further strengthened through supporting international institutions. As 'realists' are quick to point out, contradictory interests among states are played down and the question what international institutions are needed to sustain a harmonious pattern of interdependence is not really confronted. The future of interdependence can by no means be taken for granted. Rather interdependence is a manifestation of the post-Second World War order than an autonomous, unilinear process. Instead various possible world orders are now being discussed within the field of development theory, international relations theory, world system theory and international political economy.

These reorientations on the level of theory, ideology, politics and social consciousness reflected structural changes of the world economy, which can be traced back to the immediate post-Second World War period, although they were manifest only in the 1970s with the so-called crisis discussed at the beginning of this book. We noted that Gunder Frank attributed the fall of dependency to 'the crisis of the 1970s'. Hollis Chenery (1975), from a different theoretical tradition, also came to the conclusion that the world economy is 'in a state of disequilibrium not seen since the aftermath of World War II'. More than a decade later this observation still holds true. What are the main indicators of this disequilibrium and to what extent do they signify something qualitatively new?

The oil crisis of 1973 is often mentioned as the great divide. Although this event was merely one indicator of changes in the world system (the monetary disorder for instance started two years earlier), it clearly informed us about the vulnerability of Western industrialism, the finiteness of natural resources, the enormous power of the transnational companies, and the decreasing capacity of nation states to control that part of the world economy which happened to come under their exclusive jurisdiction. Not many structural changes are so visible.

The contemporary concern with 'interdependence' is to some extent a cognitive phenomenon, i.e. it is a matter of how the world is conceived. If interdependence implies a tendency towards a global social system, it may be traced far back in history. However, one could argue that this process reached a qualitatively new stage in the post-Second World War era, when many forms of human interconnectedness across the boundaries were doubling every ten years, a tendency which later was to be further reinforced through the revolution in information technology. This would not have been possible, were it not for the relative political stability of the American world order, which lasted from the end of the Second World War until the late 1960s. Since then, however, the world has, if not lacked a world order, at least tried to get along with a defective one. A world order hardly disappears completely. What happens is that significant adjustments change the way it operates.

The concept of world order covers the rules and norms regulating international economic cooperation. The rules in force during the postwar era were established at Bretton Woods (New Hampshire, USA) along with institutions such as the World Bank and the IMF. Together with GATT, which was developed later, these rules and institutions constituted the Bretton Woods postwar world order. As long as one particular world order is maintained, the rules of the game are known and widely accepted. Therefore the behaviour of the world economy, to some extent, is predictable. This is in itself a most important factor behind the integration process, which in turn reinforces the existing order (Anell 1981: 38).

Theorists of 'world society' from different theoretical approaches agree on the necessity of hegemonic power to maintain a liberal world order. As noted earlier hegemonic power is the closest the anarchic international system can come to a world government and in fact compensates for the lack of it. This means that hegemonic decline leads to a management crisis in the world economy. Due to the stable growth process after the Second World War a smooth running of the world economy came to be more or less taken for granted and incorporated with the growth paradigm. Thus it was only the decline of US hegemony which demonstrated the need for some kind of political framework for the operation of the world economy. The Bretton Woods system came to be seen as a system only when it started to disintegrate.

It is therefore less strange than it at first may seem that the phenomenon of interdependence was generally acknowledged only when the even trend towards integration was broken. Most countries are today facing the imperative of structural adjustments. It is this fact that makes the problem of interdependence so visible and concrete. The lack of predictability due to the new uncertainty about the rules of the game further adds to the problems.

World order is a superstructural phenomenon. It is therefore necessary to understand the infrastructural developments as well. On this deeper level of analysis, however, there are many question marks. The idea that the 'crisis' simply was a recession lost credibility as the 1970s passed without the return of the much sought for stable growth. On the other hand the coexistence of unemployment, inflation and substantial profits for the TNCs contradicted the classical Keynesian analysis of low-level equilibrium. Most analysts today, particularly those who take the global view, would use terms such as 'change', 'transformation' or 'transition', rather than 'recession' and 'depression'. They differ, however, on the crucial issue where we are in the transition process.

4.1.2 The world economy in transition

To make sense of the structural changes of the world economy is a demanding analytical task, but essential if we want to understand the changing preconditions for pursuing national development strategies (Bienefeld and Godfrey 1982, Kaplinsky 1984). The problem is not diminished by the lack of a consistent theoretical framework. In this context I shall make only a few empirical observations as a necessary background to the globalization of development theory, concentrating on a few key sets of issues:

- financial flows and patterns of investment;
- technological change and the internationalization of production;
- trade and other international regimes.

A convenient starting point for a discussion of the first issue is the dramatic rise in the price of oil in the early 1970s, triggering rather than causing the subsequent landslide. Surpluses from the traditional centres of finance flowed to the 'newly rich' oil producers, while highly dependent consumers, both rich (Western Europe) and poor (Africa in particular) suffered fiscal crises and foreign currency problems.

The recycling of petrodollars through commercial banks fuelled the economic transformation by providing the financial base for Third World industrialization, already in full sway in some countries. At the same time this flow of money paved the way for the debt crisis in countries where the industrialization strategy failed, or where investment was simply unproductive. In the so-called 'most affected countries' the easy credit was used up in paying the oil bill.

The result was an intensified differentiation of the Third World (Hoogvelt 1982).

A handful of states emerged as strong enough to out-compete the traditional industrial centres, but the overwhelming majority of Third World countries remained vulnerable, particularly in view of the rise of global corporate power. As economic units the TNCs dwarf most developing and even many developed countries, a striking new phenomenon in the postwar world economy. Suffice it to say that roughly 40 per cent of the total world trade now takes place between the subsidiaries and parent companies of TNCs (Corbridge 1986: 162). This gives a pattern of international economic relations not adequately accounted for in the theory of comparative advantages.

The TNCs further reinforced the changes by combining the new technology, the emerging market in industrial sites and the worldwide industrial reserve army to their advantage (Fröbel, Heinrichs and Kreye, 1980). The pattern of international investment changed, in a first phase, from the traditional industrial centres in Europe to the European periphery (in America the corresponding shift was from the East to the South – and then 'South of the border'), and, in a second phase, to any country in the Third World with a good supply of labour, political stability and other incentives, usually described as 'an appropriate investment climate'.

This reallocation of industrial investment in the 1970s depended on a number of technological developments:

- Improvements in communication and information, facilitating quick decision-making on how to take advantage of the changing global structure of comparative advantages, not only in different branches of production but later on also in terms of exchange rates (Drucker 1986).
- Improvements in transport technology reducing the importance of geographical distances.
- Improvements in technology and labour organization, making it possible to decompose complex production processes.

Thus, skilled labour in the traditional centres was replaced by unskilled labour in former peripheries, for example in Export Processing Zones, which increased dramatically during the 1970s and later was established even in mainland China. This is not to say that all Third World industrialization was of this dependent nature, but the cases of indigenous authentic industrialization are still rare. In fact this underlines the difficulty of presenting facts without an accompanying theory. The phenomenon of industrialization is not homogeneous. There are many different kinds not necessarily revealed by the statistics of industrialization, and for the purpose of interpretation both theory and disaggregation are needed.

Now to the third set of issues: trade and other international regimes. By 'international regimes' is meant 'principles, norms, rules, and decision-making procedures around which actor expectations converge' (Krasner 1985). The North generally favours market-oriented regimes, whereas the developing countries have a preference for authoritative modes of allocation, the reason being that this type of international regime provides a more stable and predictable flow of transactions, thus reducing the risk of external shocks. This NIEO approach has, as was discussed elsewhere, been unsuccessful, the commitment to market-oriented regimes has increased among significant Northern states, and there is a tendency to withdraw from international forums in which demands for regime changes are raised.

International trade is a regime where the market principle is generally adhered to – in principle. Probably the stable growth in world trade during the post-war decades (8 per cent per annum during the two decades after 1953) was unique and will not be repeated. It can be explained by an unprecedented economic growth in the OECD area, and a significant reduction in trade barriers, two conditions which underwent changes as the 1970s wore on (Kaplinsky, op. cit: 77). The 1980s were marked by the 'new protectionism', which avoids open violation of GATT rules, but nevertheless finds other ways to create 'non-tariff barriers'.

There is an obvious contradiction between the recommendations now generally given to the developing countries to opt for export-oriented strategies and the manifest lack of interest in buying the resulting goods. Many indicators suggest that the prospects for export-oriented industrialization are quite different today compared to the 1960s and 1970s, which raises the issue whether the NIC phenomenon was unique. What seems to be certain is that the world economic context facilitating this phenomenon in the 1960s and 1970s was, if not unique, at least different from the 1980s and what can be expected of the 1990s. The debt problem is still hanging over Africa and Latin America, and with that the conditionalities applied by the major financial institutions. In spite of the enthusiasm for export promotion these conditionalities have a bias against industrialization, considering that this process in reality amounts to much more than building factories: infrastructure, laying a base in the form of import-substitution industrialization, the active role of the state.

To conclude this brief overview it must be emphasized that the changes discussed under the three headings are intertwined and reinforce each other in a most complicated fashion. It also follows that many of these changes are subject to different interpretations depending on theoretical assumptions. The debate on Third World industrialization is a case in point.

4.1.3 The NIC model in retrospect

As mentioned in the previous chapter the experience of the NICs (Newly Industrialized Countries) was used as the main case for the refutation of dependency theory and held up as the model to imitate in the 1980s. It therefore deserves close scrutiny. Does it prove the effectiveness of a liberal, market-oriented open door policy (the new orthodoxy)? The literature on this issue is, for reasons easy to understand, becoming quite large. (See Cline 1982, Harris 1987, Datta 1987, Haggard 1986, Bienefeld 1987 for overviews.)

First of all – what is an NIC? We probably violate reality by using terms such as NICs or the NIC model. The concept summarizes the most recent wave of industrialization which has taken place in a limited number of countries in different regions: Latin America, South East Asia and Southern Europe.[64]

Undoubtedly much can be learned from the experiences of individual NICs, but that does not necessarily mean that they can be seen as models to imitate. We already have discouraging examples, for instance Sri Lanka, Malaysia and the Philippines. These unsuccessful efforts further underline the uniqueness of the NICs (more particularly the Asian, or the 'gang of four') as well as the great diversity of the rest of the underdeveloped world. In the case of the city states, Hongkong and Singapore, this is too obvious to deserve comment, which leaves us with two countries, Taiwan and South Korea, as far as Asia is concerned. The Latin American NICs, Brazil and Mexico, belong to a different category altogether. So do, of course, the European NICs, which, however, will be left out of this discussion.

The Latin American success stories are first of all more drawn-out historical processes with roots in the 1930s. In spite of this they differ remarkably among themselves: Brazil was authoritarian, Mexico more democratic; Argentina experienced extreme authoritarianism without accompanying development; Chile made use of dictatorship to dismantle a statist economic structure and liberate the market forces, whereas Brazil took advantage of market forces through a developmentalist state.

The cases of Taiwan and South Korea are more recent, their economic transformation more dramatic, and their social structure more egalitarian. The reasons for their successful transformation are more or less unique: their Confucian work ethic, their radical land reforms (externally forced upon them), their geopolitical importance (explaining why they had to succeed), their closeness to the Japanese giant (of whose economic expansion they formed part), and the world-economy context for their takeoff (which is not likely to be often repeated). One cannot avoid the impression that to some extent these miracles were encouraged by the capitalist world order, while countries pursuing self-reliance strategies were discouraged or even destabilized. The NICs accept and fortify the old international economic order. What they want to change is their place within it.

It seems as if the pessimism of the dependency school has been reversed into the old modernization perspective of eternal progress. A discussion of 'prototypes' v. 'emulators' in the Far East context therefore should have a sobering effect (Lamb 1981). The emulators are found in South East Asia (the ASEAN countries), and to some extent in South Asia (Sri Lanka). The Malaysian 'Look East' policy explicitly refers to Japan rather than South Korea and Taiwan (obviously for domestic political reasons), and the purpose of Sri Lanka's opening up in 1977 was to build a South Asian Singapore. Neither of these countries were successful, rather the contrary. Without going into details we could attribute this to domestic (ethnic) political factors. Secondly, the international economic conjuncture is much less favourable for the emulators compared to the prototypes. This factor has already been discussed.

If any lessons shall be drawn from the NIC phenomenon it is of little use to group them together under some fictitious label, such as 'market miracles'. None of these cases can be used as arguments for a general deregulation of the economic system. In fact the search for a key factor explaining the NIC phenomenon has been a rather futile exercise. Without such a key factor there is in fact no NIC model, only a set of variables which are different, or at least differ in relative importance, in each country.

What can be said about common denominators is possibly that their success was a matter of correct timing in switching from one development strategy to another, that the development strategy was consistent and based on a certain degree of national cohesion, and finally (and the point to be stressed in this context) that the strategy considered both internal and external constraints, as well as opportunities. The NIC phenomenon underlines the old saying 'nothing succeeds like success', but this has never been a very helpful recipe for those who tend to fail.

For the majority of countries the orthodox strategy will simply mean 'stabilization', 'readjustment' or 'restructuring' – with a 'human face' if they are lucky. In most of them even that imposes heavy strains on their political regimes, eroding their legitimacy. Few of them have sufficient social and political coherence to implement a NIC strategy, which in any case would have to depart from orthodoxy. Laissez-faire is for the very strong or (imposed upon) the very weak. Countries which successfully change the structure of comparative advantage and their place in the international division of labour (the NICs) are in between. They follow the policy of authoritarian developmentalism: to take advantage of world market forces with the help of whatever dose of interventionism they deem necessary.

4.2 Analysing world development

As noted above, there are many question marks as regards the theoretical understanding of development in the context of interdependence, now more often called world development. A whole family of 'world-system'-oriented schools have emerged. What they share is a global (instead of a statecentric) conceptualization of development. One should not expect any methodological and theoretical consensus on the interpretation of such a complex system as the 'world system'. These are converging fields of interest for both international relations theorist and development theorists (Hollist and Rosenau 1982). Though the two academic fields are becoming increasingly intertwined, we shall here keep the development aspect, rather than the power aspect, in focus.

4.2.1 The world-system approach

World-system analysis had a spectacular breakthrough in the mid 1970s but has by now also received a generous deal of criticism. The early enthusiasm has been replaced by an equally strong scepticism. I find the many condemnations overdone and partly beside the point. My position will therefore be somewhat defensive.

There is strictly speaking no such thing as world-system theory. It is rather a general approach; a theoretical project; an attempt to reconstruct an historical social science freed from a number of biases which have crippled both history and the social sciences as we have known them for the last couple of decades: biases such as evolutionism, reductionism, Eurocentrism, state-centrism, compartmentalism. Some of these biases are shared by Marxism as well, so the world-system approach is in fact being attacked both from the right and the left.

The world-system approach should be seen as a theoretical project (Taylor 1986) with the purpose of mending the weaknesses referred to. Since no single individual can master world history and contemporary events from primary sources it has to be a collective project. It is only a beginning, not a finished building, which can be dismissed out of hand as a misconstruct. Those who share the critique of the conventional social sciences as well as certain basic theoretical premises about an alternative approach form part of the project, even if their visions of the outcome – the new historical social science – may differ. It is a collective project, of which Wallerstein happens to be the most prolific writer. Even if this particular writer were to be proved wrong in most of what he writes, the project would still go on, simply because there is a need for it. A social science built on the erroneous conception of a natural history of nation states is doomed.

The origins of the world-system approach can be traced back to dependency theory, with which it shares a critical attitude toward the developmentalist framework underlying the predominant theory of the 1950s and 1960s. A second source is the Annales-school in history, which defied the positivist trend in mainstream history-writing and retained a holistic perspective.[65] Hence, holism is a basic principle in world-system thinking (Bach 1980). A third source of inspiration is the realist – or perhaps rather neorealist – tradition in international relations theory. The world-system interpretation of nation-state behaviour is thus essentially realist.

The world-system approach asserts that a capitalist world economy has been in existence since the sixteenth century. From then onwards this system incorporated a growing number of previously more or less isolated and self-sufficient societies into a complex system of functional relations (Wallerstein 1974, 1980). The process of expansion had two dimensions: geographical broadening and socioeconomic deepening. The result of this expansion was that a small number of core states transformed a huge external arena into a periphery. Between these core states and the periphery the world-system theorists identify semiperipheries which play a key role in the functioning of the system.

The core-periphery polarization implied a world division of labour in which the core countries took the role as industrial producers, whereas the peripheral areas were given the role as agricultural producers. The crucial criterion for semiperipheral, as compared to peripheral, status is thus an increase in the relative importance of industrial production. Furthermore the rising semiperipheries are strong and ambitious states, more or less aggressively competing for core status.

The world system is a social system, which, according to Wallerstein, is characterized by the fact that its dynamics are internal:

> What characterizes a social system . . . is the fact that life within it is largely self-contained and that the dynamics of its development are largely internal. (Wallerstein 1974: 347)

The expression 'largely self-contained' is, of course, somewhat imprecise. The definition is based on the following counterfactual hypothesis: If it were possible to screen off the system from all external factors, the system would still continue to function in basically the same way. We know that various parts of the world today are dependent upon each other, a situation usually described by the term interdependence. If we go back, say about a thousand years, such interdependence did not exist. Instead the world was made up of a number of relatively independent minisystems; there was no modern world system. However, history has yielded an earlier type of world system, viz. the world empire, in which the unifying factor did not consist of economic relations, but a political power-centre, which in the process of its expansion forced marginal areas under its control and demanded tribute from them.

The main difference between a world empire and a world economy (the two historically given versions of a world system) lies in the fact that the latter functions without a system of supreme political control, which proves the world economy to be a more viable and persistent structure than the world empires. As a substitute for imperial control one can see the hegemonic power exercised by a succession of dominant core states. A number of world empires have been created and destroyed, while the world economy, which has expanded steadily since the sixteenth century, today covers most of the world. In other words, it is the one and only world system, therefore also referred to as 'historical capitalism'. The rise and fall of hegemonic powers forms part of the cyclical movements of the world system, movements which basically are influenced by economic long waves. Thus the world system has periods of expansion, contraction, crisis and structural change paving the way to renewed expansion. Its uniqueness raises completely new theoretical and methodological issues in social science. It changes the prospects for social change, and thus also the strategic problem in the realm of political praxis.

The proposition that the dynamics of the system are internal implies an internalization of the external factor. Thus, the problem of the external v. the internal, which caused the dependency theorists a great deal of trouble, has, ostensibly, been solved. Like the dependentistas, Wallerstein describes the world system as capitalist but avoids the distinction between development and under-development, or central and peripheral capitalism. Thus, there is only one kind of capitalism, namely that of the world system, although its various branches may manifest themselves differently. Wallerstein thus circumvents another of the pitfalls of the dependency school, the idea of two different sorts of capitalism. In many respects the world-system approach transcends the dependency approach, but the similarities and continuities between the two should be obvious.

Wallerstein's conceptual apparatus is simple. The impact was probably due more to the suggestive way in which these concepts were used in synthetic historical analysis than to their inherent theoretical attractiveness. This work provides an impressive vision of the world's historic development, particularly during the earlier periods. With increasing interdependency and complexity of the world system the simple centre-periphery model is less helpful. In this context we are of course more interested in the world-system perspective on underdevelopment and the prospects for development.

In the world-system perspective the process of under-development started with the incorporation of a particular external area into the world system, i.e. the peripheralization. As the world system expanded, first Eastern Europe, then Latin America, Asia and Africa, in that order, were peripheralized.[66] This is a process of the world economy with its rules and its mechanisms (Hopkins and Wallerstein, 1987).

Let us choose Africa as an example of the peripheralization process. From a world-system perspective both the 'Eurocentric' and the 'Afrocentric' interpretation of Africa's history are deceptive. Both Europe and Africa became parts of a larger system and this determined their subsequent development. In other words, Wallerstein agrees with those critics who accuse him of not granting autonomy to individual actors:

> An analysis then must start from how the whole operates and of course one must determine what is the whole in a given instance. Only then may we be able to draw an interpretative sketch of the historical outlines of the political economy of contemporary Africa, which is in my view an outline of the various stages (and modes) of its involvement in this capitalist world economy.
> (Wallerstein 1976: 30)

- The first phase of the African peripheralization (i.e. incorporation with the world system) was the period between 1750 and 1900. Before that Africa and Europe were merely external arenas to each other. Trade was in 'luxury goods', the function of which was functionally unimportant to both economies. A qualitative change in their relations is suggested by the fact that the slave trade culminated around 1750 and then stayed constant for the remainder of the century. African states, which organized the slave trade and whose structures were determined by it, were peripheralized, while the regions in which the actual slave hunt took place still formed part of the external arena. The world system as such was therefore not affected by the destructive consequences of the slave trade.
- The second phase should be seen in the light of the so-called 'scramble for Africa', which eventually led to partition and the use of Africa as a production area for the core states. Different modes of exploitation led to specific forms of production in different parts of the African continent: 'rural capitalism', white

farmers' settlements, and plantations. Brought about by the peripheralization of Africa, these forms of production have since greatly influenced African political and social development.

- Also in the assumed future of Africa, i.e. the third phase, it is easy to recognize the dependency school's rather dreary outlook. The process of underdevelopment must continue to the bitter end. The core states will eventually be weakened, while the semiperiphery assumes an increasingly important role. On the other hand the position of peripheral nations, particularly the African, will deteriorate.

The conclusion is that remaining external areas, during the next fifty years, will be completely incorporated into the world system, and that the capitalist process of expansion, which started during the sixteenth century, will be completed. The inherent tendencies which Marx identified and analysed within the framework of a national, capitalist economy have constantly been countered by the existence of a periphery and an external arena. When the frontiers of the world system are reached and the world constitutes a single mode of production, the Marxist model of the capitalist system would ultimately correspond to the empirical situation.

In the present stage of the world system it is not easy to break the chains of dependency and initiate a process of self-reliant development on the national level. In fact the experiences of most Third World countries give credit to the thesis that they are part of the 'system' whether they like it or not, and that there are indeed 'limited possibilities of transformation within the capitalist world economy' (Wallerstein 1979: 66). According to the world-system theorists, development is therefore basically a matter of changing the structural position from a peripheral to a semiperipheral one, a possibility that is open to comparatively few countries. A genuine change would therefore necessitate a transformation of the world system into a socialist world government – a very distant prospect indeed!

4.2.2 Contemporary Marxism and the Third World

The world-system theorists presumably think of their framework as essentially Marxist, at least 'in spirit'. In spite of this their strongest opponents are also Marxists, who in their turn belong to different camps. This makes it rather difficult to pinpoint the Marxist theory about development in a world context. What most Marxists seem to agree on is the absence in Marx of a coherent theory of the non-European world (Banerjee 1985). The disagreements start when the foundations for such a theory are made explicit. In theoretical terms the problem is to understand the nature of non-capitalist or precapitalist modes of production. Do such attempts explode the original framework?

As orthodox Marxists I will regard those who are fundamentally faithful to the original theoretical framework developed in Das Kapital. In this way it makes sense to speak of a Marxist critique of the neo-Marxist world-system approach. Implied in this critique we would also expect to find a theoretical alternative. Certainly this alternative cannot be identical with how Marx himself conceived the world development process. Most Marxists would agree that the Eurocentric evolutionary framework that Marx shared with other nineteenth-century social scientists must be abandoned, to make sense of global development experiences of the twentieth century. As once was suggested by D. Seers, one could call those who think otherwise neoclassical Marxists (Seers 1978).

What, then, are the main differences between the world-system approach and the contemporary Marxist conception of world development? The most controversial issues seem to be the very definition of capitalism ('circulationism' v. 'productionism'), the relevance of class analysis, and the concept of mode of production. Since the theoretical differences have a paradigmatic quality, which implies that the parties conceive even the same terms differently, the Marxist critique will serve as our point of departure.

As for the meaning of 'capitalism', the world-system theorists define it as a system of exchange operating on a global level, whereas many Marxists seem to think of capitalism as a mode of production that can only be identified concretely at a national level. Thus the world system from this perspective becomes an aggregation of national capitalisms (together with other, noncapitalist, modes of production). The defining criterion of capitalism as a mode of production is the existence of alienated (or 'free') labour, and not whether the commodity production is intended for the world market or not (Laclau 1971; Brenner 1977).

This circulationist v. productionist controversy seems to be the major dividing principle between the two schools. It should, however, be noted that the difference is more subtle than the Marxist critique would have us believe. As soon as capitalism is analysed as a single world system, the exchange that takes place between various production zones cannot be equated with 'trade', and therefore Brenner's reference to 'neo-Smithianism' is less pertinent. The distinction between production and distribution may be of analytical importance within certain theoretical systems, while being without much significance in others. There is no a priori basis for making it all-important.

To define capitalism as a world system is a methodological

choice with deep-going consequences. This can be seen in the field of class analysis. Here the Marxists, not without justification, feel that the concept of class in world-system theory is more or less abandoned. To speak about a 'world bourgeoisie' and a 'world proletariat', as Amin does (Amin 1974: 24) is certainly a little premature, while Wallerstein's analysis of this particular issue is less than crystal clear, some would say confusing (see for example Wallerstein 1974: 352). It is obvious that a definition of capitalism on the world level makes the standard Marxist definition of class much too rigid, in view of the immensely different production forms which can be found within the world capitalist system. On the other hand the class concept in a world context will of necessity be rather loose. This again exemplifies that behind the struggle about concepts there is actually a paradigmatic difference. In a paradigmatic debate – if it is to be a proper dialogue – one cannot apply theoretical and conceptual standards from one paradigm to another.

Finally, the concept of mode of production is also rendered less meaningful in world-system analysis compared to mainstream Marxism, since according to the former there can only exist one mode of production, i.e. the capitalist world system. Within this Marxists (particularly those who have been exposed to the Althusserian influence) disagree strongly. In fact their alternative takes as its point of departure the coexistence of several modes of production and their 'articulation' in specific social formations. The problem of development in this perspective is the extent to which a capitalist mode dominates over other precapitalist modes, and if these other modes show a persistence, or if they are gradually undermined and dissolved.

We cannot decide here, which alternatives are more genuinely Marxist. The purpose is merely to identify approaches not deviating too much from the original Marxian conception of development. Closest to such a conception is what has been termed 'neoclassical Marxism', best represented by Bill Warren (1973 and 1980). Warren's work (interrupted by his death in 1978) can be seen as a systematic attempt to revive the original Marxian idea of capitalism (including imperialism) as historical progress, in contrast to most later Marxist theorizing about capitalism, from Lenin and onwards. Of course, the dependency literature stands out as particularly deficient in this perspective, not to speak of 'Fanonism', anarchism, utopianism and populism. This is mainstream Marxian v. counterpoint Marxist and neo-Marxist tendencies. The dualism in contemporary Marxist thinking is traced back to Lenin's *Imperialism: the highest stage of capitalism*, which implicitly concluded that capitalism in the twentieth century was devoid of positive social functions anywhere, and that imperialism had become a major obstacle to industrialization in the third World (Warren 1980: 83).[67]

Warren then went on to explain what he called the 'fiction of underdevelopment':

> The fact is that the West did not industrialize while the rest of the world stood still, nor did that industrialization retard the development of the rest of the world. Rather, the industrialization of the West from the late eighteenth century onwards (especially from the mid-nineteenth century for most countries other than Britain) tended to initiate and then accelerate modern development in the rest of the world, which otherwise would have remained comparatively stagnant. Western economic expansion aroused the non-Western world to a modernization process for which its own internal development had not yet prepared it. (ibid: 113–4)

Naturally this statement aroused a great debate about the empirical state of industrialization in the Third World, but that is not our concern here. Enough has been said to exemplify the 'neoclassical' Marxist position. Regarding future world development it holds that the growing economic interdependence must be welcomed, since it is within this context that the ties of 'dependence' are loosened and indigenous capitalisms are emerging. On this point there is a striking convergence between neoclassical economic theory and neoclassical Marxism.

Most Marxists admit that the problem of underdevelopment still remains and that this has posed theoretical difficulties.[68] One response to this problem has been to revise, modify and widen concepts used by Marx so that they can be given a somewhat broader meaning. Some cases in point are concepts such as mode of production and merchant capital, which played a more marginal and ambiguous role in the original Marxian framework. The 'mode of production' literature is particularly voluminous in French anthropology, but the logic of the approach is perhaps clearer in Laclau's early critique of Frank's much too broad definition of capitalism. The alternative he provided was based on the concept of economic system, defined as:

> . . . the mutual relations between the different sectors of the economy, or between different productive units, whether on a regional, national or world scale . . . An economic system can include as constitutive elements, different modes of production – provided always that we define it as a whole, that is, by proceeding from the element or law of motion that establishes the unity of its different manifestations. (Laclau 1971: 33)

Imperialism may, according to this view, for an extended period of time preserve archaic modes of production rather than pioneering capitalism. Thus, the problem of blocked development is acknowledged. Consequently this approach deviates too much from the original Marxian conceptions to be accepted by neoclassical Marxism (Warren 1980: 153).

A similar type of analysis is the much discussed theory of articulation, which should be seen as *a* but not necessarily *the* Marxist answer to the challenge of dependence; an answer formulated primarily by Marxist anthropology. Social anthropology has largely resisted Marxist influence, since many statements by Marx and Engels about non-European societies sound embarrassing to a discipline trying to gain respectability by breaking with its colonial heritage. Yet, anthropology must repudiate not only Eurocentric Marxism but also neo-Marxism, despite the latter's attractive 'Third Worldism'. The reason is that dependency analysis neglected the most crucial level of anthropological analysis, i.e. the local community. What happened at the local level was seen simply as a reflection of processes going on in a remote centre. Marxist-oriented anthropologists were strongly motivated in their search for a solution to both problems: i.e. the theoretical poverty of Marxism *vis-à-vis* the study of non-European societies, and the neo-Marxist tendency to deny the local level its own dynamics.

The resulting approach, called the 'theory of articulation', has its roots in the French structuralist variant of Marxism, whose foremost proponents were anthropologists like Claude Meillassoux and P. P. Rey. Subsequently British anthropologists assimilated the French debate in an attempt to go beyond the strong structural-functionalist Anglo-Saxon tradition. They also established a theoretical link between the new anthropology and the dependency school (Oxaal *et al.* 1975). One easily recognizes the dependency school's vision of capitalist expansion and penetration into the Latin American periphery, where the development is distorted and blocked. This is, however, now expressed in Marxist terminology. Underdevelopment is not explained by a mechanically applied centre-periphery model, but by the fact that the capitalist mode of production is articulated with noncapitalist modes of production. This articulation should be studied in such a way that a priori assumptions about the development process are excluded. The outcome, whether characterized as 'development' or 'underdevelopment', is attributable to differences in the type of precapitalist modes of production articulating with capitalism.[69]

In this way new questions were generated – questions which dependency theory escaped: What relation between different modes of production is expressed in the articulation? How is exploitation manifested within the class structure of a particular social formation? What is required for a capitalist mode of production to not only dominate but also replace other modes of production? Underdevelopment would, in this view, be described as a stalemate in the process of articulation. The noncapitalist modes of production have withstood the capitalist mode of production for reasons which can only be explained by a concrete analysis of individual cases.

Does this approach represent a solution to the theoretical problem of analysing underdevelopment? If so, this particular solution, as always, generates new problems. We can grasp the real manifestations of the modes of production through the approach of articulation, but on that concrete level the underlying modes have been modified precisely because of their intermixing (or 'articulation'). When a researcher claims to be studying a formation consisting of the 'pure' modes A, B and C, how to verify that we are dealing with modes A, B and C, rather than B, C and D? If, for instance, an anthropologist, after having studied some valley in the Andes, finds that this social formation (A+B+C) is dominated by mode of production A, and if another anthropologist, after a later 'restudy' finds B to be the dominant mode of production, are we then to assume that this social formation has changed, or could it be that the anthropologists had a preconceived idea about which mode of production ought to be dominant?

This gap between the abstract and the concrete levels of analysis is not easily bridged, which is shown in that the number of models of production has an alarming tendency to increase (Foster-Carter 1978). Newly discovered modes of production are being added to Marx's classical list: 'in broad outlines we can designate the Asiatic, the ancient, the feudal and the modern bourgeois modes of production as so many epochs in the progress of the economic formation of society'. Hamza Alavi and others have discussed what they call the 'colonial mode of production'. Michel Beaud has worked on the concept of a 'statist mode' (Beaud 1987). In anthropological analyses of West Africa we find the 'lineage mode'; Joel Kahn has discovered the 'petty commodity mode' in Western Sumatra (Clammer 1978: 112); M. Sahlins (1972) speaks of a 'domestic mode'; Samir Amin (1980) of 'tributary modes'; and Göran Hydén (1983) of a 'peasant mode of production'. The number of social formations 'articulating' various combinations of modes of production will be infinite, if the number of modes of production continues to grow at this rate. At the same time the world-system view that there is only one mode, a 'world mode of production', also makes the concept irrelevant. Peter Worsley concludes that Marxism would do better without 'mode of productionism', since it implies a misleading image of base and superstructure (Worsley 1984).

There are other Marxist approaches that consider capitalism as a global system without neglecting the phenomenon of blocked development. One effort of eliminating classical Marxism's inherent theoretical difficulties with this problem was to elaborate on Marx's own distinction between industrial capital and merchant capital (Kay 1975). The basic difference between the two is that only industrial capital is capable of generating surplus value, whereas merchant capital must be created by means of nonequivalent, or 'unequal' exchange.

According to Marx, merchant capital was a historical prerequisite to the growth of the capitalist mode of production, subordinated under industrial capital which, in contradistinction to merchant capital, should be seen as a social relation. In precapitalist society the merchant obtained his profits by organizing trade between different societies, a trade entirely different from exchange between regions integrated into the same system of production and reproduction (Weeks and Dore 1979: 82). Only under such circumstances will there be capital accumulation in the Marxist sense.

Merchant capital may of course expand by exploiting divergencies in the cost of production between different economic systems (unequal exchange), or as a result of a deterioration of the producers' situation caused by various kinds of political oppression (colonialism), but it will never revolutionize the production relations and create capitalism. The precapitalist mode of production was instead strengthened by political alliances between the comprador bourgeoisie and precapitalist ruling elites. According to Geoffrey Kay:

> Capital created underdevelopment not because it exploited the underdeveloped world but because it did not exploit it enough. (Kay 1975: x)

However, to see merchant capital as the cause of underdevelopment is a simplification, and in retrospect the approach must be considered a dead end. The 'trade', if it can be called that, between centre and periphery was fundamentally different from the classic, long-distance trade between precapitalistic societies. Furthermore, the former type of trade required extensive changes in the relations of production in the periphery. Kay's claim that the industrial capital left the carrying out of these changes to the merchant capital is therefore not just another doubtful widening of Marx's definition of merchant capital (which was limited to the sphere of circulation), but 'a shattering of it' (Bernstein 1977: 57). This is the old distinction between central and peripheral capitalism re-emerging in the guise of a distinction between industrial capital and commercial capital, and with the same inherent theoretical difficulties.

One Marxist response to the dependency challenge thus has been to extend and modify the meaning of Marxist concepts like mode of production and merchant capital. These 'solutions', nevertheless, led to new problems. There is still a considerable amount of theoretical work to be done before the gap between the Eurocentric Marxist tradition and the social reality in the Third World is bridged (Chilcote, 1982, 1984). In the later part of the 1970s the relevance of the Marxist approach was strongly asserted.

> The only theory able to inform the struggles of the proletariat and other exploited classes throughout the world is that of historical materialism, which is itself the site of a continuous struggle to maintain its integrity and hence its effectiveness. (Bernstein 1979: 97)

4.2.3 Neostructuralism

We have earlier referred to the structuralist criticism of the neoclassical view of development as a smooth, gradual, equilibrating process. In Latin America the early debate concerned the problem of inflation. It has been shown that this structural tradition was close to, even part of, the CEPAL variety of dependency analysis. It has been fighting with the back against the wall during the wave of orthodoxy in the late 1970s and early 1980s, but has also found a new task in criticizing the stabilization programmes (Sutton 1985). It is possible to distinguish a 'neostructural' school with roots in earlier structuralism and modified by dependency analysis.

According to this approach the world economy constitutes a structured whole, and its constituent parts display various forms and degrees of dependency. Dependence is thus a structural property characterizing all units of the world economy in varying degrees. According to D. Seers this would be the proper basis for a classification of national economies (Seers 1979a). Even a very rough classification, taking relative dependence on import of oil, food and technology as its point of departure, immediately reveals a pattern which is rather different from the conventional centre-periphery model. Among the 'least dependent countries' we find the United States, the Soviet Union and China; among the 'semidependent countries' Japan, East Germany and Nigeria; and among 'dependent countries' Brazil, Cuba and Portugal.

The modern neostructural approach includes many problems and levels of analysis, as is evident from the following quotation:

> This approach (i.e. the 'structural-internationalist' models') views underdevelopment in terms of international and domestic power relationships, institutional and structural economic rigidities, and the resulting proliferation of dual economies and dual societies both within and among the nations of the world. Structuralist theories tend to emphasize external and internal 'institutional' constraints on economic development. Emphasis is placed on policies needed to eradicate poverty, to provide more diversified employment opportunities and to reduce income inequalities. These and other egalitarian objectives are to be achieved within the context of a growing economy, but economic growth per se is not given the exalted status accorded to it by the linear stages model. (Todaro 1977: 51)

This approach can be applied to different levels of society. In this context we are primarily concerned with the world level analysis. A good example of this is provided by Osvaldo Sunkel's and Edmundo Fuenzalida's research on what they call the transnational capitalist system (Sunkel and Fuenzalida, 1979).

In a way this approach can be regarded as dualism at a global level, since the most striking feature of the system is the polarized development of transnationalization on the one hand and national disintegration on the other. To start with the first aspect, the capitalist system has changed from an international to a transnational structure of a remarkable consistency, and with the transnational corporations as the most significant actors. Of particular interest is the emphasis on culture, usually rather neglected in structural analysis. It is hypothesized that a new transnational community is emerging, made up of people from different nations, but with similar values and ideas, as well as patterns of behaviour.

If we then turn to the other side of this dual global structure, the national societies as a consequence of the transnationalization process undergo a process of disintegration, implying a disruption of indigenous economic societies and a concentration of property and income. This process of marginalization, in turn, explains the tendency toward repression and authoritarianism which can be seen in both underdeveloped and developed countries. However, at the same time the national societies are generating a variety of counter processes that assert national and/or subnational values, sometimes reactionary, sometimes progressive.

According to Sunkel and Fuenzalida the transnationalization thesis may be a useful conceptual framework in several respects. First, it suggests why growth in many poorer countries should be associated with growing inequality. Secondly, the combination of transnational integration and national disintegration offers a more subtle concept of dependency than most earlier models, as the impact of dependency on the internal structure of the periphery is emphasized. Thirdly, the analysis of the way in which this differential impact cuts across the usual categories of class seems especially important. Parts of the bourgeoisie, the petty bourgeoisie, the industrial working class, etc, are integrated into the transnational system, while other parts are not. This demands a refinement of class analysis.

There are, however, certain limitations in most of these global approaches, for example excessive structuralism and economic reductionism. Obviously there is more to development than changing positions within an international division of labour, and the production and distribution of material goods. Development concerns people, it affects their way of life and is influenced by their conceptions of the good life, as determined by their cultures. In the chapter that follows we will deal more with the content of development than its structural characteristics, but first the implications of the new global approaches for national development strategies must be considered.

4.3 Development strategies and the world system

Development theory has so far evinced a national bias, i.e. the very idea of development has been closely associated with the nation state. Is the fact of interdependence now so imperative that it is no longer possible to talk about 'development' for a unit smaller than the world system? What, in this case, are the implications for development strategy? And what shall we mean by that often used but rarely defined concept?

A development strategy is evidently a strategy for promoting a process of development. Thus it contains two components: ends (development) and means (strategy). The two can be discussed separately, but usually they are mixed up (see Fig. 4.1)

Fig. 4.1 The relation between development ideology and development strategy

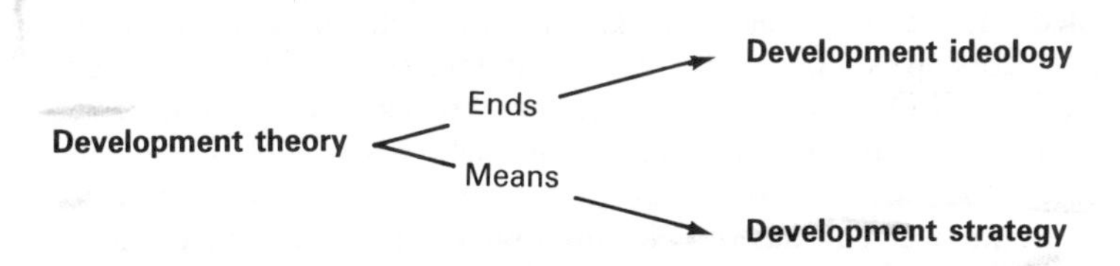

Modern development theory, in contradistinction to classical evolutionary theory, was from the start both normative and instrumental. This implies that:

- Theorists allowed themselves to have opinions about what development ought to be – and, as could be expected, the opinions often differed.
- It was assumed that development was a process that could be controlled and steered by actors, normally the state.

This explains why development has been a contested concept, and development theory an area of conflicting schools.

So far we have touched upon the ends component, or the normative aspect, of the modern concept of development. The means component, or the instrumental aspect, relates to development strategy, i.e. the means to promote development, or what the strategists in any particular case conceive as development.

Thus, development strategy is essentially an empirical concept immediately related to state behaviour. It can be seen as the more or less explicit plan which a nation state applies to the problem of reproducing itself in material terms, in relation to its own population and resources, and in relation to the external world. As we earlier noted, development strategy forms part of nation-building.

As students of development theory we can allow ourselves to be normative, because development obviously has to do with values. As students of development strategies, however, we should observe this phenomenon as it actually appears in the real world. Development strategies are what relevant actors (so far mainly the state) do to promote what they define as development. The concept implies a capacity to act. Without such capacity the concept has no meaning. This is why a crisis for the nation state also implies a crisis for development strategy.

4.3.1 Dilemmas of global reformism

To the extent that it still makes sense to talk about national development paths, they cannot, as within the classic modernization paradigm, be conceived of as endogenous processes. The theoretical interpretation of global development depends on how one conceives the empirical phenomenon of interdependence. Both NIEO and the subsequent proposals of the Brandt Commission are here referred to as examples of global reformism, since they conceived the world as a single system and, as a consequence, emphasized the imperative of change for the system as a whole. The major problem with this reformist strategy is what agents of change can be identified, since the whole conception of purposive intervention, implied in development strategy, is so strongly connected with the state as the dominant actor. In the previous chapter we related the NIEO demands to the breakthrough of the dependency perspective which began to form part of the UN conference jargon during the early 1970s. The concept of self-reliance was redefined as 'collective self-reliance' as an expression of Third World solidarity. However, the emphasis was not so much on self-reliance as on a fair deal for the South on the world market. To the more radical dependency theorists this was a case of conservative reabsorption.

NIEO thus was a political rather than an economic strategy, aiming at establishing a trade regime based on authoritative allocation.[70] The economic demands include price stabilization, indexation, increased aid volumes, changes in the monetary system, technology transfer and food security on the lines of a rather conventional view of development as modernization. On the other hand it was silent on ecological balance, internal social reforms and basic human needs. Even if some of its formulations suggested a self-reliant approach, consistent with the dependency paradigm, the main proposals in fact implied a path to development through more trade with the industrialized countries and access to their technology, rather than creating conditions for the development of independent technological capabilities (Villamil 1977: 90).

Among the industrialized countries the US was particularly reluctant to accept the NIEO demands, since they implied authoritative (political) rather than market allocation. Three strategies could be used to delay or contain such reforms. First, the unilateral strategy (project independence), secondly the alleviationist strategy, e.g. quick actions on minor problems but no concessions on basic demands, and, thirdly, the acquiescence but delay strategy, e.g. symbolic declarations in anticipation of the erosion of the solidarity of the Group of 77 (Ries 1977: 78–9). The third strategy was particularly effective since, as soon was realized, the long process of negotiations on a large number of technical issues could protect the rich world for a long time to come (Hveem 1976). This was the strategy opted for by the European countries, as they were, and still are, much more dependent on Third World resources than the US.

The European countries preferred to conceive NIEO in terms of increasing trade and widening markets, which may serve the purpose of stimulating the world economy and take it out of the depression, increasingly evident during the 1970s. To accept the more radical demands would have been to accept a moral responsibility for the poverty in the Third World, i.e. the logic of the dependency paradigm. Instead, the leading development agencies in the 'centre' argued that the principal cause of underdevelopment was not external but internal. This view was implied in the 'redistribution with growth' strategy, and may be a not unimportant factor explaining its popularity in the North soon after the ideological breakthrough of NIEO. While there was a growing consensus among Third World countries about the need for a radical reform of the international economic order, the idea that radical domestic reforms in the poor world were called for was equally fast gaining strength among development agencies in the developed countries. Thus, the World Bank could dismiss NIEO as elite demands, while assuming the role as protector of the rural poor. However, the basic problem with NIEO, as with all global strategies, was that it was a strategy without obvious implementing actors.

If we turn to a later document outlining an international development strategy, the Brandt Commission report North–South: A Programme for Survival (1980), the problem remains basically the same. The Brandt proposals were, as we noted, based on the concept of interdependence. The North–South dialogue, that had been initiated with the NIEO declaration, very soon came to a deadlock. One reason was that the rich, but stagnate industrial countries could not afford to satisfy the demands raised in the NIEO document. In order to avoid a process of polarization, the rich countries preached the theme of interdependence and common destiny of all mankind. This philosophy permeates the Brandt Commission report.

In terms of development strategy the report articulates a Keynesian solution to world poverty by proposing a Massive Resource Transfer (MRT). According to this theory the poor peoples of the world are to function as the unemployed in Keynes's system. As they make use of the financial resources to buy goods produced by the industrial countries, the economic problems of the latter would be solved as well. The rich and poor countries are to move forward together rather than the poor countries being given benefits at the expense of the rich world, which was the strategy of the NIEO and the earlier UNCTAD reform proposals. Thus, whereas NIEO can be seen as a product of the milder CEPAL variety of dependency theory, the Brandt Commission exposed a 'global Keynesianism'. This proposal was, paradoxically, aired at the same time as neoliberal monetarist economic policies were being tried out on the domestic scene in many North Atlantic countries, for example Great Britain.

Naturally the responses to this proposal varied in accordance with different development ideologies.[71] Its radical liberalism is certainly not accepted by the new right, urging the underdeveloped countries to balance their budgets, liberalize their economies and find out their comparative advantages. This alternative, subsequently termed Global Reaganomics, completely overshadowed the Brandt report during the Cancun meeting in October 1981.

The criticism from the left, on the other hand, questioned the idea of mutual interest between North and South, implied in the interdependence thesis. In the leftist view the integration of the Third World into a system of global interdependence is more likely to increase conflict than stability. The 'strategy of survival' is in fact a strategy for the survival of capitalism.

For the proponents of Another Development, finally, the report is not sufficiently aware of the ecological impact of global capitalism and the institutional difficulties of raising world production catering for basic needs.

A radical structuralist interpretation of 'interdependence' leads to very different conclusions with regard to development strategy, which can be illustrated with the world-system approach.

If the possibility of breaking with global capitalism and the establishment of national socialist systems was implied in much theorizing on dependency (although this option never was fully elaborated in terms of development strategy) this development path is bluntly ruled out in the world-system perspective. According to Wallerstein we are now living in the transition from capitalism to socialism, but 'it will undoubtedly take a good 100–150 years yet to complete it' (Wallerstein 1980: 179). There exist within the framework of the world economy socialist movements controlling certain state machineries but no socialist national economies (1974: 351). No reforms, however radical, can establish 'socialism in one country' because the 'free' world is not prepared to let it happen. The world-system perspective is rather extreme on this point, but it can nevertheless be concluded that the globalization of development casts doubts on the viability of strategies focusing on national development. In the following section we shall look at some concrete examples.

4.3.2 Experiments with delinking

One reason for the shift of emphasis from national to global processes was the bad record of dissociative development strategies in the 1970s. In what follows we shall discuss some cases of self-reliant development and the problems these countries – Tanzania, Ghana, Jamaica, Nicaragua – had to cope with.

In terms of power base a policy of self-reliance implies that the regime in question depends on its own people rather than getting support from external powers. One illuminating case of a country finding it hard to solve the dilemma of mobilization from below v. mobilization from above is Tanzania. Her attempts at self-reliance did not create the same international controversy due to security policy reasons as some other cases to be discussed below. Neither was there a strong and politically organized indigenous bourgeoisie that felt challenged by rather modest socialist measures.

The Arusha Declaration of 1967, an early example of the strategy of self-reliance, should be seen as a result of the frustrations of a very poor country with social ambitions but rather unattractive to international capital. Its 'failure' must be seen in this perspective and any critical assessment should consider what the alternatives were at the time when the new development strategy was chosen. The strategy worked rather well until the early 1970s, especially when compared to other sub-Saharan economies (Bienefeld 1982).

Since then Tanzania has gone from crisis to crisis, but so have most African countries, regardless of what particular development strategy they pursued. One particular problem in Tanzania, however, was that the 'ujamaa' part of the strategy of self-reliance was de-emphasized after 1973, and instead a massive villagization campaign, to a large extent based on coercion, was initiated. The mixed economy approach was replaced with the statist approach, which in Tanzania meant party rather than government control (Hartmann 1988). As the process of mobilization gradually became 'from above' rather than 'from below', and the role of force became more prominent, one basic component of the strategy was lost.

In the case of Tanzania the general African problem of foreign exchange constraints due to expensive imports and sluggish growth of agricultural exports coincided with an increasingly ambitious strategy of self-reliance, defined as to include also basic industries (the Basic Industrial Strategy). This was a change in the definition of self-reliance, compared to Nyerere's original conception, and the inspiration came from the dependency school. The new strategy turned out to be unrealistic, if not in itself, so at least in the context of the world depression, combined with international inflation in the 1970s. Without foreign assistance it would have collapsed earlier. If the Tanzanian government becomes dependent on the IMF, while at the same time having lost the confidence of the peasant population, it may not be easy to revive even the original and much more modest policy of self-reliance. At present quite different policies are tried but the split between party and government makes the future uncertain.

Another African country where a strategy of self-reliance has been tried is Ghana. Since Ghana became independent in 1957 three civilian republics have been destroyed through military interventions, and over the years the span of civilian rule became shorter (Hettne 1985). Ghana's current crisis was discussed in Chapter 1.

As far as the problem of dependence is concerned one civilian (Nkrumah) and two military regimes (Acheampong and Rawlings) at least at the policy level entertained explicit ambitions to make Ghana more self-reliant. Nkrumah, like Nyerere after him, stressed industrial development within a general scheme of socialist transformation; Acheampong put emphasis on agriculture of a large-scale type; and Rawling's current programme is best described as an externally supported basic-needs-cum-stabilization strategy.

As far as the issue of performance is concerned it is of importance to keep the time perspective in mind, as well as the relationship between ambitions and outcomes. When assessing the development performance of the various regimes one must consider the objective conditions for development in every single case. Nkrumah was obviously highly ambitious and he also started out from a rather solid financial situation of which not much was left in 1966. All governments after that were faced with increasingly difficult economic prospects. In acute economic crises – and there have been many of them in Ghana – people in the short run suffered irrespective of what economic policy the incumbent government held on to. Most often these crises were caused by a drop in the cocoa price.

Since there has been an accumulation of problems, Rawlings must compare badly to any previous regime, simply because of the depth of the economic crisis which more or less imposes 'self-reliance' of some sort on the country. Considering these circumstances it does not make sense to compare regime performance using quantitative indicators of development. For years to come the development strategy will have to be a strategy for national survival. The main contradiction is caused by Ghana's firm restructuring programme, devised by the IMF, and the populist image of Rawlings' regime, which created expectations about a better life for a suffering population. Thus the internal power base is constantly eroded.

If self-reliance implies structural transformation of the economic system and social relations within a country pursuing a reasonably ambitious version of the strategy, this is obviously also the case at the level of external relations. A change in the pattern of international relations would in most cases have security policy implications within a particular region. The more sensitive the region is, the more serious these implications would be. Since self-reliance means increased autonomy in relation to international market forces and capital interests, it could, with some justification, be interpreted as hostility *vis-à-vis* the 'free world'.

This 'hostility' typically leads to countermeasures from those external interests that have been challenged. To the extent that the countermeasures appear in a more or less systematic and planned way, we can say that the country trying to be self-reliant is subject to a campaign of destabilization. Let us briefly consider one sensitive region, the Caribbean and Central America, and two cases, Jamaica and Nicaragua (Hettne 1983).

In Jamaica 'the Puerto Rican Model' v. 'the Cuban Model' have been subject to a lot of dispute and of course these two models have been related to both domestic and foreign policies. Before 1972 there was little difference between the political parties: Jamaica Labour Party (JLP) and People's National Party (PNP). In 1972 the PNP took over, and, under the leadership of Michael Manley, its programme was radicalized. In the field of foreign policy, it was clearly stated that Jamaica belonged to the Third World and that Cuba was to be a close ally. This was in great contrast to the 1960s when Jamaica followed the Puerto Rican way, implying an open and export-oriented economy and encouragement of foreign investment, and with a foreign policy orientation towards the United States. Thus, the reorientations of both development policy and foreign policy during the 1970s were taken as unfriendly acts by the United States, and when the economic strategy ran into difficulties, the IMF was not particularly helpful. In fact it was regarded by the Manley government as the agent of destabilization. IMF was a major issue in the 1980 election, and so was the 'Cuban connection' in the JLP campaign against Manley.

Jamaica was thus gradually pushed into a severe economic, social and political crisis by a combination of the inherent adjustment problems of a new development strategy and the destabilizing policies to which foreign capital interests tend to resort when a dependent nation embarks upon a 'hostile' political course. This was accompanied by increased political violence, reaching a level unprecedented in Jamaica's political history. It was generally believed that import of guns was part of the destabilization campaign and that the CIA was involved. The government was deeply shaken, and found itself in a difficult, political crisis. The majority of Jamaicans now were reluctant to continue on the road of self-reliance, particularly as there were contradictory views even within the government. The 1980 election put an end to Manley's rule. The return to power of the JLP led to fundamental shifts in Jamaica's policies of development and foreign affairs during the 1980s.

The Nicaraguan case is in many ways different from that of Jamaica. The challenge represented by Nicaragua is greater and the military component of the destabilization process has become the main factor. In Jamaica there was a threat of civil war in the late 1970s, whereas Nicaragua is threatened by full-scale war. Most Central American states in fact conceive Nicaragua as a threat by merely providing an example of an alternative development strategy and an alternative model of society. This shows that the choice of development strategy has political implications both internally and externally, which will react back on the conditions and possibilities of pursuing a certain course.

Of course the idea of self-reliance was rather utopian in the Nicaraguan case, since the economy was traditionally oriented towards the North American market and the country was dependent on food imports. In spite of this the Sandinista government was bent on an economic policy that implied structural changes, challenging both external and internal interests. In other words, the Sandinistas were asking for trouble. Destabilization by overt military operations was increasingly resorted to, when it became clear that the nonmilitary destabilization campaign did not produce much result. That the nonmilitary form of destabilization failed to achieve its ultimate purpose does, however, not mean that it was harmless. The economic destabilization, for example, has created many problems. The economic problems will turn into social problems, which subsequently become political. The only possible answer to a policy of destabilization is political mobilization. Ideological consciousness is Nicaragua's hard currency. With a 'devalued political currency' (i.e. a legitimacy deficit) there are small chances for the country to survive. For this reason one could consider political devaluation as the essence of destabilization. There are many ways to achieve this.

The examples of Jamaica and Nicaragua seem to show that the strategy of self-reliance does not necessarily fail simply due to lack of economic realism, even if this also may be part of the picture. History shows that in strategically sensitive areas the policy of self-reliance challenges superpower hegemony. Countries such as Tanzania and, more recently, Ghana have, as was discussed above, also made attempts at increasing their degree of self-reliance without much external (as distinct from internal) political controversy.

The cases of Nicaragua and Jamaica are different. They illuminate the problems involved in choosing a strategy of development that works against the interests of both external and internal political forces. The differences between the two cases are also significant. The Nicaraguan revolution has so far survived an increasingly hostile destabilization campaign, ranging from the cultural to the military arena, and the reason for the survival is the level of political mobilization and ideological consciousness. These preconditions were lacking in Jamaica and therefore a less intensive destabilization campaign had more far-reaching political consequences. Consequently, in Nicaragua, the USA escalated the campaign from destabilization to covert military intervention. On the whole the option of self-reliance became much more difficult in the era of global militarization which started in the late 1970s and continued through most of the 1980s.

To sum up: The new interest in global theories can be regarded as an effort to go beyond dependency, and to create a framework in which both centre and periphery, as well as the relations between them, are considered. In the current development debate there seems to be an overreaction to the weaknesses of the dependency school and pessimistic determinism as far as the strategy of self-reliance is concerned. Export-oriented industrialization strategies, successfully carried out by some of the so-called 'newly industrialized countries' (NICs), are recommended instead. This change in the theoretical debate on development should, like the new Cold War, be related to the changes in the world economy during the 1970s. The decline of the US and the ensuing rivalry between industrialized countries, in which Japan was particularly successful, create unique possibilities for certain countries with a reasonably advanced industrial base, cheap labour and stable (repressive) regimes. On the other hand, the scope for self-reliance and other more or less radical strategies was reduced due both to economic and to political reasons. The economic reasons were changes in the terms of trade: rising import prices for energy, food and capital goods; falling prices for many of those raw materials that still dominate many underdeveloped economies. Of the political reasons the most important was a felt need for streamlining the allies and consolidating control over the spheres of interest of the two superpowers. Here the USA possesses a much more varied repertoire of instruments for economic destabilization, whereas the Soviet Union to a larger degree must rely on diplomatic and military means.

This is not to underestimate the internal problems of mismanagement, isolation and repression, which sometimes accompanies self-reliance strategies. The 'Burmese way to socialism' is a recent case in point.

In conclusion, we would like to stress that the 'failure' of self-reliance must be understood in relation to structural and political changes in the world and should not only be explained by inbuilt weaknesses of national development strategies. These global changes, as expressed in the 'new' Cold War, have made it even more difficult to implement strategies of self-reliance. For social, political and cultural reasons only a limited number of countries can follow the NIC strategy. A number of countries, such as the Philippines, Malaysia, Sri Lanka, Chile (after Allende), Jamaica (after Manley) and Puerto Rico, learnt the difficulties of implementing the NIC strategy. Thus, the relevance of self-reliance (as a strategy rather than as a nationalist ideology), implied in the dependency approach, should not be judged only by the setbacks of this strategy in the 1970s. Rather they should be seen as learning experiences.

4.3.3 Between the Scylla of autarchy and the Charybdis of integration

Radical delinking has now been ruled out as more or less impracticable by all camps. On the other hand the original arguments against full integration with the world economy have not lost their relevance, simply because the remedies proved too simplistic. Neither the closed nor the open strategy constitute the blueprint of development. Development theory therefore faces a new challenge:

> What are the prospects of development in the currently less developed economies, and to what extent do these prospects depend on the effective implementation of national development strategies which take cognizance of the dominant trends in the international economy?
>
> (Bienefeld and Godfrey 1982: 1)

The underdeveloped economies are an extremely heterogeneous category. Therefore no general theoretical formulations of their problems, and no universal cures can be valid for the category as such. That such a statement perhaps seems self-evident today actually shows how far we have travelled since Rostow's stages. What development theory will have to be preoccupied with from the point of view of a modified dependency approach is to define in a precise way the actual relationship between different national situations and a continuously changing international context, and what room for manoeuvring there is for national development strategies, as both external and internal constraints are taken into consideration (Seers 1981). This awareness is basically what was missing in what we now could call the classical or 'vulgar' dependencia approach.

Between the most general statements of classical dependencia and particularistic studies of unique cases, there is the level of middle range theorizing and typologies. There is need for more systematic knowledge about both the evolution and dynamics of the world economy and the domestic political foundations for different types of development strategies, if the crucial relationship between the external and internal developmental factors shall be grasped. If we have learnt anything from the dependency debate, it would be that any development strategy must be based on a careful assessment of its broader political and economic context. The way a specific country links up with this larger context depends on the prevailing power relations which constitute a particular regime as well as the conjunctural stage of the larger system, which of course also is in a constant process of transformation.

The integration v. autonomy option is the crucial dimension behind current classifications of development strategies (see Fig. 4.2). It has been a key issue in development theory over the years, beginning with the Listian critique of British political economy, or as Friedrich List himself put it: national v. cosmopolitic political economy (List 1841).

This issue – whether there is a contradiction between national and international development – lay, as we remember, behind the rise of development economics. The 'national political economy' was carried further by the dependentistas who argued in favour of a

Fig. 4.2 A classification of development strategies

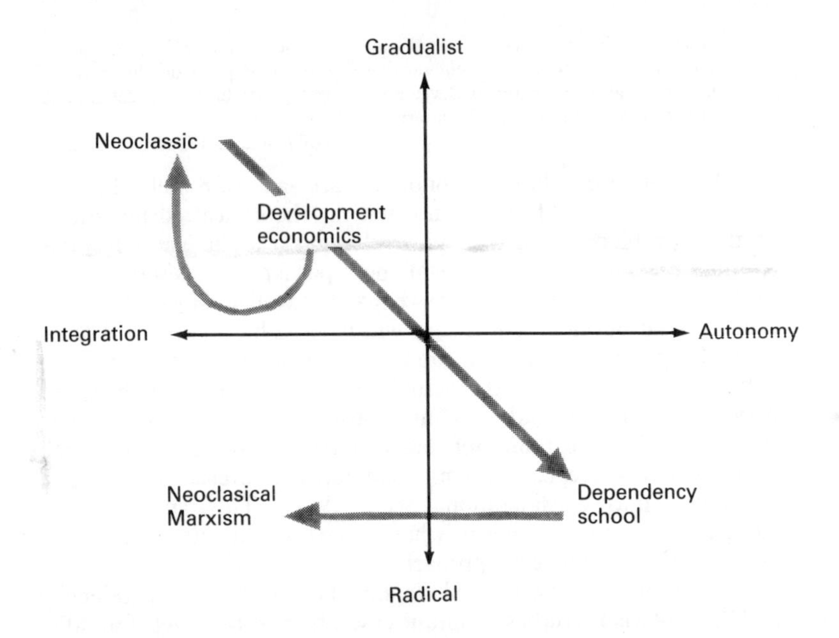

radical delinking from the world market. The pendulum has swung back since then, but the original issue has not lost its relevance. The shifting positions are shown in the figure above.

One way out of the impasse in which development theory finds itself, and a means to revitalize the now dormant field of development studies, would therefore be to focus on the comparative study of development strategies and internal and external constraints on their implementation. For this a good typology of development strategies is much needed. It can be constructed in many ways, in a more or less systematic fashion, or in an *ad hoc* approach informed by current experiences of development. Keith Griffin for instance, identifies six development strategies (Griffin 1988):

- Monetarism – which assumes the long run efficiency of market signals in resource allocation, although introduced in a period of crisis with the short run purpose of economic stabilization. The (economic) role of the state is minimized.
- Open economy – which places special focus on policies promoting foreign trade and other external links as engines of growth. It is perfectly compatible with an active supply-side-oriented state.

- Industrialization – which emphasizes the manufacturing sector as the leading source of growth, oriented towards the domestic or foreign market (or combinations). State intervention is normal.
- Green Revolution – which gives priority to increased productivity and technological (rather than institutional) change in the agricultural sector as a means to foster overall growth.
- Redistributive strategies – which start from redistribution of income and wealth and a high degree of participation as a means to mobilize people in the development process.
- Socialist strategies – which above all stress the role of the state sector in development: planning, state farms, public manufacturing enterprises, although the role of the central state may vary from extreme statism to self-management.

We must not take it for granted that all countries pursue distinct development strategies. On the contrary, as Griffin emphasizes, most countries do not follow any identifiable strategy, and certainly not for long. This is increasingly the case, due to the weakening of the state in the Third World, and also due to the world-economy crisis. Thus the role of development strategies for many countries today tends towards crisis management rather than socioeconomic transformation, which of course significantly reduces the relevance of development theory.

A different approach to the comparative study of development strategies is to base the classification more firmly on a theoretical perspective, using dimensions derived from it. World-system-oriented analyses strongly underline the limited possibilities for a nationally controlled, autocentred development process. This, by the way, seems to be an emerging consensus among leftists and rightists. Of course, conventional wisdom has always had it that every country should make use of its comparative advantages. This is now aggressively reiterated by the 'counter-revolutionaries'. Mainstream Marxism, similarly, accused the dependentistas of cultivating a myth of autocentred development that never really existed anywhere. The world-system theorists have much in common with the dependencia tradition, with the significant difference that they warn against 'delinking', pointing to the many miserable failures ocurring in the 1970s.

> Everywhere, the reality has been that the fact that a movement proclaims the unlinking of a state's productive processes from the integrated world economy has never in fact accomplished the unlinking. It may have accomplished temporary withdrawal which, by strengthening internal production and political structures, enabled the state to improve its relative position in the world economy. (Wallerstein 1980: 183)

According to world-system theory, there are in principle only three strategies: the strategy of seizing the chance, the strategy of promotion by invitation and the strategy of self-reliance (Wallerstein 1979: 76). The first is a classical one, roughly identical with what was earlier discussed as state capitalism. It basically involves aggressive state action to tranform the structure of comparative advantages with the purpose of capturing external markets. Of course not all countries have the capacity of seizing the chance. The second, development by invitation, is based on existing comparative advantages, such as low wage level and a general hospitability (cf. the liberal model). Obviously not all countries will be found hospitable enough. The third, self-reliance, deviates from both the other strategies by its inward orientation. In the context of the present world system, however, it is the strategy which is most unlikely to succeed according to world-system development thinking.

The present world system is, as we recall, considered to be in a crisis, and, as we also remember, crisis implies both risk and opportunity. Thus, crisis means a period of transition. The option of 'seizing the chance' is a strategy which systematically exploits the anarchical conditions of the transition. At the same time it contributes further to the anarchy by constituting a challenge to the old core, and the old economic order. The countries trying to pursue strategies of socialist self-reliance – from Chile to Grenada – were of course challengers to this order in a more radical sense – and were dealt with accordingly. The NICs, in contrast, have been political friends even if they, in some respects at least, behave as economic foes.

The recent development record seemingly supports the NIC strategy, in comparison with the unsuccessful delinking strategy. The limitations of the NIC strategy is, as Gunder Frank puts it, 'intuitively clear', but the fact that delinking is not only a policy attempted by progressive governments but also a weapon used against them in the form of destabilization 'gives cause for reflection about the rational utility of delinking in the world today' (Frank 1983b).

The question of the optimal mix of autonomy and integration is by no means a new one. It therefore makes a lot of sense to compare historical development strategies from the point of view of their way of tackling the external environment. As Dieter Senghaas has demonstrated, the European experience is significant from this perspective (Senghaas 1985). His classification of European development strategies departs from the question how the more or less successful countries dealt with the autonomy v. integration issue in the contemporary world context, England's overwhelming economic dominance:

- dissociative development based on the dynamics of the domestic market (Germany);

- associative export-led development (Switzerland);
- associative-dissociative development (the Scandinavian Model);
- dissociative state-capitalist development (Russia);
- dissociative state-socialist development (the Soviet Union).

The order of the cases is roughly chronological. One conclusion to be drawn is therefore that delayed development became increasingly difficult to achieve and that it depended on a strong interventionist state (op. cit: 38). This conclusion was also drawn by dependency theorists. One underlying dimension behind this particular classification is the issue of inward v. outward-oriented development, or what Senghaas calls dissociation v. association. One striking point is that very few European countries actually developed in accordance with the way the World Bank and the IMF now recommend the underdeveloped countries to develop. Rather they followed, in varying degrees, the old Listian formula.

In a posthumously published book Seers (1983) combines the internal/external dimension (which he calls nationalist v. antinationalist) with a second dimension, based on the degree of egalitarianism. By combining these two dimensions, four different ideological positions are identified: socialist and liberal varieties of internationalism, arguing for open-door development strategies; and radical and conservative varieties of self-reliance and delinking (see Fig. 4.3). As Seers points out, development policy is basically a balancing act. What he calls 'the room to manoeuvre' differs objectively between countries as well as historical situations, but also subjectively between various observers: 'bureaucrats, for example, tend to exclude all possible policies outside a narrow range, whereas many academics assume policy to be largely or totally unconstrained' (op. cit: 56). To succeed in development is to make use of the available room to manoeuvre in order to accumulate, rationalize the national production system, and steer the country into the appropriate niche in the world division of labour. Is this what the NICs are doing?

I submit that this is precisely what they are doing. We have already discussed the limited applicability of the NIC model to the majority of still underdeveloped Third World countries. Its most important message is not to liberalize the national economy, but rather to exploit the liberalization of the external world through a consistent and forceful mercantilistic economic policy. The secret behind success is a strong government, committed to development, ruling over an obedient people, and relatively autonomous *vis-à-vis* influential social groups – a very rare situation in the Third World today.

Samuel Huntington, a well-known realist, once said: 'It is not the type of but degree of government that matters.' Similarly,

Fig. 4.3 A classification of development strategies according to Dudley Seers (1983)

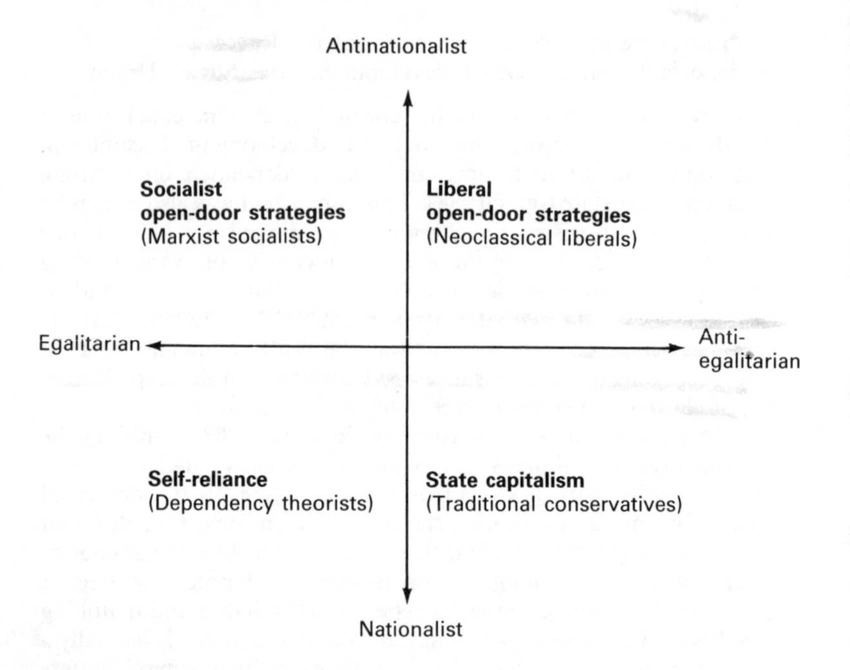

Gunnar Myrdal long ago stressed the distinction between 'hard' and 'soft' states. These insights have not lost their relevance, although they for different reasons were neglected. The NICs and the phenomenon of 'authoritarian developmentalism' again have brought our attention to this problem, easier to define than to do anything about.

The NICs did not choose between import-substitution industrialization and export-orientated industrialization. They did both, and firmly shifted emphasis at the right moment. This is the crucial test for a developmentalist regime, since any type of development strategy will develop its own vested interests opposing any changes that hurt these interests. The problem is not to choose a strategy but to pursue it consistently and to change it when the situation so demands. The necessities are often defined by external conditions over which no single regime, whatever its domestic strength, has full control. The international context is, as was noted above, constantly changing, creating obstacles as well as new possibilities. For this reason it is dubious to draw strategic conclusions from one decade and expect them to be valid in another. For the same reason development theory must be flexible and responsive to the concrete development strategies applied to constantly changing situations. It is the fate of the theorists to lag behind the strategists, but it is the role of theorists to accumulate wisdom on which the strategist can draw – in a critical and selective way.

Chapter 5

Dimensions of Another Development

> The modern industrial system, with all its intellectual sophistication, consumes the very basis on which it has been erected.
>
> E. F. SCHUMACHER

This chapter is devoted to normative approaches in development theory, i.e. contributions which deal with development not in terms of how it actually takes place but rather how it should take place. The distinction between positive and normative, 'is' v. 'ought', is often implicit and this has certainly contributed to some intellectual confusion in this field.

The early concerns of development economics to a large extent reflected the interest of the ruling elites in developing countries. The modernization paradigm took it for granted that the societies characterized by industrial capitalism are universally desired, but in fact no people ever voted for capital accumulation and industrialization, processes that have usually implied a substantial amount of coercion. The more backward a country is, the more coercion seems to be needed. The modernizing elites have, wisely enough, been silent on the horrors of creative destruction, or what Peter Berger (1977) calls the 'pyramids of sacrifice'. There is, thus, no way of escaping value-judgements in development theory.

A widely circulated example of the normative approach from the mid 1970s is the Cocoyoc Declaration adopted by the participants in a UNEP-UNCTAD symposium at Cocoyoc, Mexico (1974). The subject of this meeting was resources and development and the principal opinion of the participants was that mankind's predicament is rooted primarily in economic and social structures and behaviour within and between countries. It was declared that a process of growth that did not lead to the fulfilment of basic human needs was a travesty of development. However, development implied more than basic needs, e.g. freedom of expression, self-realization in work, etc. Furthermore, there was a need for the rich to reconsider 'overconsumptive' types of development, violating the 'inner limits' of man and 'outer limits' of nature. Thus was the 'alternative' trend in development theory born.

Theories that are explicitly normative are no less important than positive analyses, since visions of the good society influence actual development to the extent that development is affected by political actions and human will. The utopian trend in development theory is perhaps best summarized in the concept of 'Another Development' popularized by the 1975 Dag Hammarskjöld report What Now, prepared on the occasion of the Seventh Special Session of the United Nations' General Assembly, and further elaborated in the journal Development Dialogue. Sometimes the concept 'alternative development' is preferred, as in the dossiers of the International Foundation for Development Alternatives (IFDA). The journal Alternatives, published by the Institute for World Order (New York) and the Centre for the Study of Developing Societies (Delhi), should also be mentioned in this context.

The main importance of these normative approaches is that they focus on the content of development rather than the form. Economic growth models show the predominance of form over content in the early Eurocentric phase of development theory. The Marxist preoccupation with the development of the productive forces reveals a similar bias in what in those days was conceived of as the rival approach. Dependency theory, albeit more explicitly normative, did not really consider the purpose and meaning of development. In fact the more radical alternative approach has come from neopopulist theorists in the centre, although some of them have been inspired by Gandhian philosophy and Buddhist economics. These theoretical trends are compatible with and to some extent even constitute a counterpart to the emergence of green politics in the Western world.

According to the alternative development thinking popularized by the Dag Hammarskjöld Foundation and the magazine Development Dialogue, Another Development should be defined as:

- Need-oriented (being geared to meeting human needs, both material and nonmaterial).
- Endogenous (stemming from the heart of each society, which defines in sovereignty its values and the vision of its future).
- Self-reliant (implying that each society relies primarily on its own strength and resources in terms of its members' energies and its natural and cultural environment).
- Ecologically sound (utilizing rationally the resources of the biosphere in full awareness of the potential of local ecosystems

as well as the global and local outer limits imposed on present and future generations).

- Based on structural transformation (so as to realize the conditions of self-management and participation in decision-making by all those affected by it, from the rural or urban community to the world as a whole, without which the goals above could not be achieved) (Nerfin 1977: 10).

Structural transformation is, of course, not unique for the alternative approach, but the orientation of this transformation is indicated by the normative content of the other four principles. It should also be noted that the idea of endogenous development was central to the modernization paradigm as well. However, in the alternative approach there is no universal path to development. Every society must find its own strategy in accordance with its own needs.

In what follows we shall first analyse the alternative approach from a metatheoretical perspective in order to grasp its significance and avoid a premature dismissal of it as utopianism. Even if it is true what Samir Amin has said (Amin, 1985) about the green currents – that they are symptoms of rather than solutions to the crisis – they will become more and more relevant as the crisis deepens. Therefore, we shall try to put these currents in a historical and sociological perspective and also relate them to political praxis. After that we shall make an assessment of the contributions of alternative development theory in terms of the following four major themes: egalitarian development, self-reliant development, ecodevelopment and ethnodevelopment.

5.1 The sociology and politics of anotherness

From a positivist point of view, the concept of Another Development is nothing but development ideology, without relevance for understanding development as it actually takes place in the present world. Whatever development is occurring today (for example in the NICs) is in fact more consistent with the conventional modernization paradigm than with the set of ideals implied in what we, for lack of a better term, call Anotherness. Why, then, this sudden interest, particularly within the rich countries, in Another Development?

One way of explaining this ideological trend would be to look upon it as a phenomenon of overdevelopment rather than underdevelopment. In the Third World the interest in Another

Development has a different background. The quest for alternative solutions and a 'third way' to development was never rooted in any mass consciousness but typically formed part of a usually rather superficial ideology of the modern elite, rejecting both North American and Soviet materialism for nationalist reasons. If socialism was referred to in this context, it was not the scientific socialism of Marx and Engels but African or Arab socialism, bent upon revival and preservation rather than creative destruction and modernization (Berger 1977: 60). This ideology seems to have been more common in Asia and Africa than in Latin America, and stronger among the first generation of Asian and African leaders (Nasser, Nehru, Sukarno, etc.) than among the politicians of today, who, for pragmatic reasons, usually opt for mainstream solutions.

Therefore, the paradox that remains to be explained is why the concepts of Another Development, implying small-scale solutions, ecological concerns, popular participation and the establishment of community, etc., have met with relatively more enthusiasm in the rich countries, while they to a large extent are being rejected in the poor.

The reasons for the latter part of the paradox are not far to seek. Small may be beautiful, but it does not entail power (as far as the ruling elite is concerned). The masses in the Third World will never reach the material standard of living at present maintained in the West (and by Third World elites), but some urban middle classes in some areas may, at least theoretically, achieve this. Consequently those chosen to become 'modern' do not intend to be fooled into some populist cul-de-sac.

Why then this interest for Another Development in the North? My answer to this would be that the collective consciousness of the industrially advanced countries is going through a process of transformation. Western Development thinking can be analysed as a dialectical process between a Mainstream (or dominant) development paradigm and its Counterpoint. In terms of this framework the Counterpoint is now gaining ground, whereas the spokesmen for the Mainstream have a hard time finding a way out of the present impasse; a solution consistent with the world-view of automatic growth and eternal progress.

5.1.1 The Counterpoint reasserts itself

Perhaps the clearest manifestation of the Counterpoint is the populist tradition. This tradition is rooted in the 'Gemeinschaft' type of society, whereas the dominant thinking in the West rationalizes the 'Gesellschaft' model. The classical form is of course the Russian narodnik movement, but populism can, as a more general concept, be seen as a worldwide phenomenon, articulating the interests and values of peasant societies which are threatened by the penetration of industrial capitalism. Today ideas and values in many ways strikingly similar to classical populism, but born in a quite different context, are re-emerging. Below the various historical manifestations of the Counterpoint referred to here are briefly summarized:

Conservative romanticism was a reaction against the emerging nineteenth-century bourgeois society. In terms of social class it was of course articulating the interests of a threatened traditional elite. However, the criticism of industrialism, in spite of being an ideology (in the Marxian sense), had a universal character transcending the more immediate situation of class struggle because of the wider relevance of this criticism. It is that particular dimension of conservatism which is of interest here, not the European counter-revolution in general. In daily parlance conservatism means being against change but it would be equally appropriate to think of conservatism as preserving what is valuable from the point of view of certain values. When the planet is threatened we all tend to become conservative.

To simplify a little it was particularly the negative aesthetic and ethical implications of the ongoing social change that were stressed by the romanticists and, associated with these, the problem of 'alienation'. This problem was to reappear in Marx's early writings, and later in the works of Weber, Freud, Morris, Fromm and Marcuse.

The transition from 'Gemeinschaft' to 'Gesellschaft' was taken to be painful for the individual, who was alienated from a context in which a development of his or her whole personality was at least objectively possible, to a situation where certain aspects of this personality were grotesquely overdeveloped, whereas others were suppressed. To Marx alienation was a necessary consequence of the capitalist mode of production. Weber pointed out that the irreversible rationalization of modern society made it dull and unbearable because it lost its 'charm' (Entzauberung).

Weber was basically pessimistic about the outcome of this process, whereas Marx could think of a future mode of production in which the problem of alienation was solved and the full potential of the productive forces nevertheless was taken advantage of. This is what distinguished Marx from the conservatives. He took over their problematique, but looked for revolutionary solutions. Since no communist Utopia yet has been created, this dilemma has not lost its relevance.

The Weberian pessimism, on the other hand, finds its more recent echo in this statement by Robert Nisbet:

> Faith or even interest in progress is hardly to be expected in a civilization where more and more groups are ravaged by boredom; boredom with world, state, society, and self. (Nisbet 1980: 349)

Nisbet even concludes that 'never in history have periods of culture such as our own lasted for very long'. This is conservatism in the true sense of the word.

Utopian socialism was a complex and at the same time rather inconsistent ideological trend which only partly can be treated as a manifestation of the Counterpoint. It is conventionally recognized as a distinct ideology, mainly due to the fact that Marx and Engels described the works of the three major proponents (Saint-Simon, Owen, Fourier) as examples of 'prescientific socialism' – all of them reacted against large-scale industrialism and the process of proletarianization. In spite of the fact that these utopian socialists had strikingly parallel careers, they had a rather low opinion of each other's work. One could regard Saint-Simon as fairly 'Mainstream' because of his association with Comte's ideas, his firm belief in Progress and his worship of *le système industriel*. Owen's commitment to small-scale organization makes him a more typical representative of the Counterpoint, but on the other hand he was in fact an industrialist with a strong belief in Progress. Fourier, who strongly criticized the dullness of industrial society and stressed the importance of 'passions', comes closer to the Counterpoint. It is significant that his contribution is the only one to have experienced a revival in our time (Taylor 1982). Thus, both utopian socialism and anarchism preferred a decentralized society, but whereas the former tradition was regulative, the latter's conception of the ideal society was more spontaneous and improvised. We can conceive of this difference as two distinct approaches to decentralization.

The regulative or planned approach necessitates either an overall control over social change, which the utopian socialists never were able to attain, or the creation of many microsocieties, which together were strong enough to influence the general direction of social change. Such attempts were also made, but the contradictions between these microsocieties and the macroprocess became unsolvable and the social experiments succumbed to the logic of superindustrialism. This is a dilemma also faced by contemporary alternative communities – probably to an even larger extent, since they often depend on the social security system forming part of the supposedly unsustainable modern industrial system.

Anarchism was above all a reaction against statism in its various dimensions and, among those, enforced industrialism in particular. Like the romanticists, the anarchists favoured a decentralized and multifaceted social structure which made individual self-realization possible. Like the utopian socialists the anarchists were, to say the least, a mixed lot, many of them rather eccentric. In fact some anarchist traditions were very close to early radical liberalism, whereas others tended to be more collectivistic. Of course this had much to do with specific national patterns of culture, Russian anarchism for instance tending towards anarcho-communism, American towards anarcho-capitalism. As a political movement anarchism was traditionally strong in southern Europe, particularly in Spain where the Civil War, as well as what followed after it, put an end to this peculiar legacy. However, as indicated by Woodcock, in referring to the neoanarchism of the 1960s, the Counterpoint never dies:

> It [anarchism] can flourish when circumstances are favourable and then like the desert plant lie dormant for seasons and even for years, waiting for the rains that will make it burgeon. (Woodcock 1975: 453)

Modern anarchism of course retains its hatred of the centralized state but has also discovered ecology (Bookchin 1980, 1987). Ecologism in contradistinction to environmentalism stands out as a new synthesis of traditional anarchist thinking about the necessity for decentralized social, economic and political organization.

Populism has for long been a contested concept (Ionesco and Gellner 1969) which tends to create confusion due to its very different connotations in different political contexts:

- Reactions of an agrarian more or less self-reliant society against capitalist penetration (the narodnik movement in Russia, Third World populism).
- A developmentalist alliance between the national bourgeoisie and labour unions (Latin American populism).
- A small-scale farmer capitalism fighting against finance capital and urban interests (North American populism).
- Antibureaucratic petty bourgeois movements in stagnant welfare states.

Of these varieties the three first may express ideas and values which are more or less close to the Counterpoint, but I would regard the Russian populism as the classical case (Hettne 1976). It is an extremely well-articulated populist ideology, mainly because it attracted a number of brilliant Russian intellectuals in the late nineteenth century.

The narodniks argued against industrialism as a large-scale and centralized form of production (i.e. the state-capitalist strategy) but did not oppose all kinds of technological progress. Rather the Russian 'privilege of backwardness' should have been taken advantage of, i.e. 'to strive for what the others have already achieved not instinctively but consciously [. . .] knowing what should be avoided on the way' (Vorontsov, quoted from Walicki 1969: 116).

The narodniks were also antistatists, which is natural in view of the repressive form Russian industrialization took. Their views on the state however contained many nuances (Berlin 1979: 217). Perhaps their most interesting contribution was the criticism of the idea of division of labour. They would not accept the sacrifices in terms of human personality (recognized but considered necessary by both Adam Smith and Emile Durkheim) to get a more differentiated complex and efficient society. Mikhailovski's law of progress is very different from Mainstream ideas on development.[72] It is the Counterpoint:

> Progress is the gradual approach to the integral individual, to the fullest possible and the most diversified division of labour among man's organs and the least possible division of labour among men. Everything that diminishes the heterogeneity of its members is moral, just, reasonable, and beneficial.
>
> (Walicki 1969: 53)

The green ideology can be seen as a contemporary synthesis of neopopulist and neoanarchist ideas that were revived in the 1960s, forming part of the new left movement in the US and Europe, and later to be merged with ecology and peace movements. These ideas bear a certain resemblance to the classical populism and anarchism in urging for community ('Gemeinschaft') and in their distaste of industrial civilization.

As the green movements transform themselves to parties (or exploit party politics as one arena of struggle) there is a need for a new 'green' economics which, as Paul Ekins points out (Ekins 1988), stands out as a counterpoint to the assumptions of formal economics:

- Much formal economic activity is harmful and much is the wasteful result of having to remedy those harms.
- The most important work in society is done outside the formal economy.
- Money can be a useful economic means but is a totally inappropriate economic end.
- The local economy matters far more to people than the national aggregate.

The green critics of industrial society are of middle class origin and it may be as true of them as what Marx once said about the utopian socialists:

> They all want the impossible, namely the conditions of bourgeois existence without the necessary consequences of those conditions.
>
> (Quoted from Kitching 1982: 32)

This is true of all the historical manifestations of the Counterpoint discussed above. The antisystemic articulations of protest are typically parochial, reflecting specific historical class interests. The criticism as such may nevertheless be more universally valid, transcending the various class positions and informing later protest movements in the continuous resistance against an inhuman system.

In fact, the old Russian populism, contemporary Third World populism and the present upsurge of green movements in the West may be seen as an example of intellectual interaction between 'developed' and 'developing' societies. It is most significant that there exists one intellectual current which is rooted both in Western and non-Western traditions, and that this type of development thinking is drawing on contributions from both Western and Third World thinkers. The green perspective has the potential of bridging the North–South dualism in development thinking and becoming truly transnational. Since market exchange is the dominant mode of economic integration on the world level it is not unreasonable to expect world-wide antisystemic movements in support of local communities to appear (Friberg and Hettne 1985).

In the case of the Counterpoint I cannot point to established visible structures such as industrialism, urbanism, professionalism, militarism and other Mainstream structures. Since the Counterpoint is an emerging phenomenon the supporting structures are also emerging – only partly visible and partly understood. Some relevant structures that may function as future bastions for the Counterpoint are:

- the community;
- the informal sector;
- Third System politics.

5.1.2 The rise of reciprocity

Let us now turn from the level of ideas, worldviews and ideologies to the level of institutions, more specifically the institutionalization of economic life.

It is obvious that a certain mode of institutionalization corresponds to a certain set of ideas, a social paradigm. The Mainstream development paradigm is rooted in an institutional structure with manifestations such as industrialism, statism and professionalism, whereas the Counterpoint has expressed itself mainly on the ideological level. This is the weakness of its contemporary manifestations. It lacks an institutional base from which 'modernity' can be opposed. The inherent attractiveness of the Counterpoint does not provide a sufficient base for social change in the desired direction. Only in the wake of traumatic failures of Mainstream development – 'development catastrophes' in the poor world (the Bophals to come) and technological setbacks in the rich (the 'Challengers' and 'Chernobyls' to come) – will provide the pedagogical showcases for rethinking.

Mainstream social science is subordinated under the 'Zweckrationalität' of the current institutionalization of society. Theories based on the 'Wertrationalität' of human survival are consequently dismissed as nonscientific and too value-loaded to be accepted in the academic world, which has its own mode of institutionalization, recruitment, socialization, negative sanctions and rewards. For this reason we cannot expect universities to be the intellectual base for the production of new knowledge. Mainstream economics, for instance, is an expression of the 'Zweckrationalität' of market exchange, and the major alternative, Marxist theory, is historically linked to state intervention and the command economy.

Any 'third approach', 'green' economics in contradistinction to 'blue' and 'red', is hard to discover in contemporary social science for the simple reason that it hardly exists in the real world of modern economic life. Positive social science is overwhelmingly concerned with what *is*, not with what *ought* to be. This is a dilemma for economists trying to develop a New Economics (Ekins 1986).

The present always predominates over the past and the future in the way we try to understand society. The way out of this bias is to marry history and social science, as has been discovered and rediscovered repeatedly. A major contribution in this respect was that of Karl Polanyi (whose disciplinary abode therefore is impossible to specify). I shall here draw mainly on some of his ideas in my attempt to relate the Mainstream-Counterpoint contradiction to the institutional level. The fundamental problem is to understand the substantive character of market exchange and what possible substitutes there are in the institutionalization of economic life, in the future as well as in the past (Polanyi 1957, 1977).

The concept of 'market' has two meanings, one concrete, namely the market place, another abstract, referring to the market system. Market places are a more or less universal phenomenon as we can learn from history and anthropology. They all operate in accordance with the same basic logic, regardless of how the society at large has chosen to institutionalize economic life. The prices of those goods that are exchanged on the market fluctuate according to supply and demand conditions and determine 'profits' of different commodities in the short run and resource allocation between production of different commodities ('investment') in the long run.

The crucial point made by Polanyi is that societies which are completely dominated by the market principle, implying that also land, capital and labour have been commodified, is a recent phenomenon. Historically there has been two other possible economic integration mechanisms: reciprocity and redistribution.

The former refers to the socially embedded forms of exchange in small-scale symmetric communities; the latter to politically determined distribution in stratified societies marked by a centre-periphery structure. Both modes of distribution were undermined by the growth of market exchange. However, as the market principle penetrated all spheres of human activity, thereby eroding social structures, redistribution had to be reinvented in order to provide people with the necessary social protection. This was the origin of the welfare state. Thus modern industrial societies are typically distinguished by a market-redistribution mix. Depending on the nature of this mix we call some 'capitalist' and other 'socialist'. In neither system does reciprocity play any role in economic transactions outside the family.

In the present economic crisis in the First World there are those who tend to rely on redistribution and the public sector (the Keynesians), whereas others look to the market principle for guidance (the neoliberals). These are the two Mainstream solutions. There is, however, also a Counterpoint solution stressing the role of symmetric exchange and cooperation at the community level. This solution (if it turns out to be one) will therefore have to rely on the revitalization of the reciprocity principle inherent in the phenomenon called informal economy.

At present there are clear signs that the legal and illegal informal economy is growing in importance. An article in Newsweek (30 June, 1986) reports that underground economic transactions represent at least 10 per cent of the GNP in Europe. Dimensions and forms vary. In Italy the underground economy involves more than three million people. To a large extent it is due to various attempts at surviving in a situation of economic stagnation, but there are also examples of more positive and ideologically conscious efforts to apply the principle of reciprocity in new systems of production.

This changing role of the three distribution principles is suggested in Fig. 5.1. The figure is purely heuristic and not based on any calculations of the actual importance of the different principles. We must also guard ourselves against evolutionary and deterministic implications, which were carefully avoided by Polanyi.

In the long historical perspective the market principle has gradually assumed more importance at the cost of the two competing principles, particularly redistribution. This mode at an

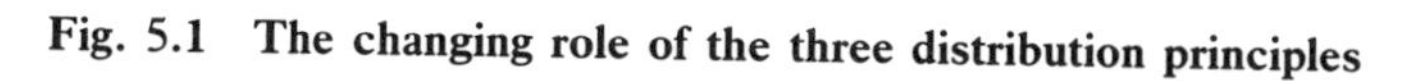

Fig. 5.1 The changing role of the three distribution principles

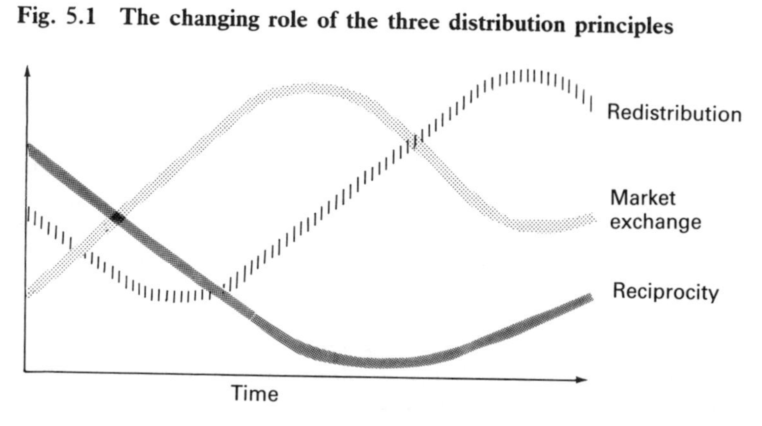

early stage overtook the principle of reciprocity which lost its relevance as soon as larger political formations emerged. This redistribution-market contradiction can also be seen as a competition between the polity and the economy. The typical situation in earlier economic history seems to have been 'politics in command'. Today one can speak of an 'economic revenge'.

The redistributive empires of Egypt and Mesopotamia have even been described as forms of 'state socialism', and whatever trade there was must be described as administered trade rather than market trade (Polanyi, Arensberg, Pearson 1957). Even in antiquity, during which period a capitalistic 'takeoff' according to some economic historians was close, political control was the rule. After imperial disintegration in Europe the three principles may be said to be relatively balanced, but as capitalism replaced feudalism – 'the great transformation' in Polanyi's terms – the predominance of the market principle was a fact. This was – again in the words of Polanyi – 'the origins of our time'.

Capitalism meant an enormous strengthening of both state and market at the cost of the autonomy of local communities. In spite of the state – market contradiction – the ideological controversy of our time – both these institutions were intimately related during the rise of capitalism. Under mercantilism national markets were created through the subordination of local economic systems under absolute state power. At the same time the foundation was laid for a world market controlled from Europe and the new nation states, who typically became colonial states. The strongest of them all was Great Britain, hegemonic enough to allow the market principle to determine economic transactions on the world level and to force other states to apply free trade principles, at least until the crisis of the 1870s.

The decades before this crisis, as well as the decades before the crisis of the 1930s, were periods of more or less complete market dominance. Economic Man was thus born in an institutional setting where men and women themselves were commodified. The market society was a utopia in the making, but, as happens with utopias, reality overtook it in the form of recurrent crises. Politics took command in the form of fascism, communism and welfareism (social-democratic reforms based on Keynesian economics). In the postwar world welfareism was established in most of Western Europe and communism in Eastern Europe as two varieties of the Mainstream paradigm. Both systems are today in crisis. The green current is one of the responses.

These trends, the future relevance of which we can only speculate about, are no longer confined to the ideological level. There is a structural counterpart at the more concrete socio-economic level in the re-emergence of informal economy (Gershuny 1979). Our understanding of that not really new phenomenon is only in a very early stage. There are several synonyms or partly overlapping concepts: unofficial economy, underground economy, parallel economy, black economy, hidden economy, informal sector, etc. The proliferation of concepts indicates a phenomenon that is relatively unexplored and it is uncertain if the concepts always refer to the same phenomenon. It exists in both developed and underdeveloped societies, but there must obviously be some crucial differences involved here. What makes it possible to deal with them in the same context is simply that they are responses to crises in the three 'worlds', in the system of market exchange as well as the system of redistribution.

In a Third World context the distinction formal/informal was first used in the 1972 ILO Report on Kenya. No less than seven criteria were used: easy as opposed to difficult entry to the economic system; reliance on indigenous rather than foreign resources; family in contrast to corporate ownership; small as against large scale of operations; labour-intensive and adapted versus capital-intensive and imported technologies; informally rather than formally acquired skills; and unregulated and competitive, as opposed to protected, markets.

This approach is difficult to translate into operational terms, so in practice simplified dichotomies such as small – large, self-employment – wage employment, registered – unregistered, etc. are employed. The problem with those is that the two resulting sectors due to the variety of transitional forms between the two poles are not qualitatively distinct (Sandbrook, 1982, p. 16).

In the industrialized world ('first' and 'second') informal economy has even more complex connotations. In a capitalist market system 'informal' can imply also production, distribution and consumption networks that emerge outside or at least on the periphery of the market nexus. In planned socialist economies informal would rather refer to the nonregulated economic activities that usually take the form of market exchange, but nota bene in the substantive rather than the formal sense of the word, exchange in the market place not in the market system.

A system of market places is qualitatively different from a market system and perfectly compatible with the more balanced pattern of integrative principles which must be the solution to the present crisis. However, the institutionalization of new economic patterns will hardly take place without political interventions. The state no longer has the key, as was the case in the previous crisis, neither has it the legitimacy. It nevertheless remains a potential actor in putting social limits to market exchange and permitting the emerging reciprocity structures to operate, even if this means stepping down from its role of omnipotence.

The self-regulating market was a political arrangement. The maintenance of this system on the world level has presupposed strong, hegemonic states. In order to protect less developed societies against the market the state grew stronger and stronger, until the societies needed protection not only from the market but also from the state. The state must be distinguished from the society, and the re-embedding of the economy in the society presupposes not only a reduced role for the market but also for the state. This implies the rise of reciprocity. It also implies a new form of politics.

5.1.3 Third System politics

The concept of Third System has been used both by the World Policy Institute and the International Foundation for Development Alternatives (IFDA), but in a slightly different way. Both see the Third System as a system of power represented by people acting individually or collectively through voluntary institutions and associations. It is the main bearer of new values and visions and thus the most important source of change. The First System is a system of power comprised by the governing structures of territorial states, i.e. the state system. However, for IFDA the Second System is associated with economic power; the market and market forces, such as corporations and banks. The World Policy Institute includes these economic actors in the First System, while intergovermental organizations – for instance the United Nations – are put into the Second System. In the IFDA terminology the international (interstate) organizations form part of the First System (Falk 1980).

Since the interstate system is an expression of the logic of the nation state we prefer to follow the IFDA terminology suggested by

Marc Nerfin.[73] This also gives us the possibility of distinguishing more clearly between economic power (the transnationals) and political power (the states), and we see people's power as a third kind, based on consciousness, organization and action. To simplify, one could talk about three main actors: the Prince (First System), the Merchant (Second System) and the Citizen (Third System). Thus in this context citizens constitute the Third System when they do not seek either governmental or economic power.

What, then, does the Citizen struggle for? What is the essence of Third System politics? Basically it is a defence of the autonomous power, originally held by the people, against encroachments both from the First and Second Systems (or the Prince and the Merchant). Historically these encroachments were associated with state-formation and the growth of capitalism. In the process, accommodation was sometimes found between the interests of Prince, Merchant and Citizen, in spite of many objective conflicts. Today, however, the Prince and the Merchant have grown into monsters, which by their transnational economic and military power threaten the very existence of the Citizen. This threat even includes extinction by nuclear war. We can therefore expect conflicts between various incompatible social projects emanating from the three systems, particularly in the fields of development, peace and environment.

Third System politics is related more to the territorial than to the functional principle, although the political issues may be translocal. For example the environmental issue is often concretized and dramatized in the local space, but the sources as well as the solutions must be looked for in larger, even transnational systems. Functional macro-organizations have largely been failures from the point of view of the poor, while more often being instrumental for emerging elites, challenging old elites and trying to get access to state power. For this reason and because the problems of the less privileged groups cannot be compartmentalized according to the functional principle, Third System politics tend to have a territorial base.

Third System politics in the Third World is even more issue-oriented than in the industrial world and the links between different antisystemic movements even within one single country are still weak and will remain so for a long time. Therefore it is somewhat speculative to say that this struggle can be seen as a direct support for the Alternative Model. The support is rather of an indirect kind, since the countermovements and the emerging nonparty political formations counter the modernization process and give a concrete content to alternative patterns of development.

One observation to be made is that actors in the Second and Third Systems are transcending the nation state in pursuing their respective projects, in the Third System the ecological and peace movements in particular. A new form of violation of state laws by activists has already become front page news in the last couple of years as a consequence of the emerging transnational identifications. Legal measures against law violations stemming from the new transnational or global consciousness will become increasingly difficult to justify, i.e. they will not appear as legitimate by a growing part of the population.

5.2 Egalitarian development

No development strategy explicitly aims for inegalitarian development but conventional strategies implicitly assume inequality (social or regional) as a necessary price for growth. In contrast egalitarian strategies give a higher priority to redistribution than to growth. One example is the Basic Needs Approach (BNA). In the early 1970s it was widely agreed that economic growth did not necessarily eliminate poverty. Rather the economic growth that actually took place in most developing countries went together with increases in absolute poverty. In response to this the BNA favoured a direct approach, i.e. a straight relationship between development strategy and elimination of poverty rather than waiting for the 'trickling down' effects of growth (Emmerij 1988).

5.2.1 Emergence of the basic needs approach

Tracing the intellectual history of BNA is rather demanding since the idea as such is self-evident – to say the least. After all, it was the poverty of the postcolonial world that made development a theoretical and practical concern. Many countries where political conditions made economic planning and development policies 'people-oriented' of course tried to implement Basic Needs strategies without necessarily using that terminology (Green 1978).

The new thing about the basic needs debate in the mid-1970s was obviously the distinction between economic growth and needs satisfaction (or indirect v. direct approaches to poverty eradication). Secondly, the strategy was adopted by several international agencies such as ILO, UNEP and the IBRD which of course made it important to scrutinize the concept of Basic Needs and the strategies informed by it. Obviously these strategies differed in radicalism due to the context in which they were discussed and implemented (Sandbrook 1982). Most important was undoubtedly the ILO's World Employment Programme, launched in 1969, and the 'Declaration of Principles and Programme of Action for a Basic Needs Strategy of Development' adopted by the participants in the 1976 conference on World Employment.

Central to the ILO strategy of basic needs was creation of employment which was given a higher priority than economic growth *per se*. Of course no contradiction between growth and employment was assumed but on the other hand it was not taken for granted that growth necessarily led to increasing employment either. In order to maximize employment more emphasis should be put on agriculture and the informal sector, but employment creation must also lead to a rise in real incomes.

The World Bank adopted a modified version, in which redistribution and growth were treated as complementary rather than contradictory elements of a new development strategy. During the latter part of the 1970s the BNA thus reached a wide acceptance ranging from World Bank economists, who enriched their growth models with social indicators, to global reformers, who inflated the BNA into a future programme not only for the 'underdeveloped' but also for the 'overdeveloped'. The adoption of the BNA by so many different theoretical traditions and interests was bound to affect the very content of the concept.

5.2.2 The interpretation of human needs

The diverse background of the various elements constituting the BNA and the resultant complexity of the concept explains some of the confusion that has characterized the debate on the value of the approach as well as the concept as such. Thus a first distinction should be made between the discussion centred on the viability of the strategy and the more philosophical debate on human needs, partly stimulated by the strategy discussion.

In what follows we will be more concerned with the general basic needs (or human needs) discussion. Here a distinction has been made between, on the one hand, a universal and objective interpretation of 'needs' and, on the other, an interpretation that is subjective and historically relative (Lederer 1980). To this then comes the distinction more basic – less basic (or basic – nonbasic) already implied in the concept of BNA and therefore easily forgotten.

To make priorities among different needs is highly problematic, however, and therefore there are those who want to delete 'basic' from the needs debate (ibid: 6). One argument for this is the difficulty in finding criteria for deciding on the relative importance of needs. On the other hand one would think that the practical value of the whole approach is tied to the possibility of making priorities.

Returning to the even more complicated distinction between objective and subjective, the former school defines human needs as something that applies to all human beings and that could be quantified and measured; the latter takes human needs to be historically relative and therefore to be seen in the context of specific social systems. The first school refers to those needs that in all societies are necessary for physical reproduction, whereas the second approach has more to do with what makes life worth living in different cultures. In the first view basic needs are quantifiable and a universally valid definition is possible. According to the second view basic needs is a qualitative concept that partly falls in the realm of philosophy and religion. It covers also transcendental values and is relative with respect to different cultures. A universal definition is therefore impossible. As G. Rist points out:

> It is not enough to answer that one must eat, it is also necessary to take into account what this person will eat, how he will eat it, and with whom he will eat. (Lederer 1980: 237)

The first approach belongs to the positive, the second to the normative approach in development theory. The former is closely related to the redistribution with growth strategy of the World Bank during the 1970s, one important difference being that it implied channelling particular resources to particular people (Streeten 1979). In this context of Another Development we are primarily concerned with the second approach but, as will become clear, the two interpretations are not easily separated in practice.

As Benjamin Higgins points out (Higgins 1980) these two uses of the concept can be found already in the basic needs 'bible' (ILO 1976). Below follows what seems to be the positive definition:

> Basic Needs, as understood in this report, includes two elements. First, they include certain minimum requirements of a family for private consumption: adequate food, shelter and clothing are obviously included, as would be certain household equipment and furniture. Second, they include essential services provided by and for the community at large, such as safe drinking water, sanitation, public transport, and health and educational facilities.

But already in the Programme of Action the concept has grown more normative and considerably more complex:

> It is important to recognize that the concept of Basic Needs is a country-specific and dynamic concept . . . In no circumstances should it be taken to mean the minimum necessary for subsistence; it should be placed within a context of national independence, the dignity of individuals and peoples and their freedom to chart their destiny without hindrance.

The Basic Needs concept thus conforms to a general tendency that can be noted in development theory; key concepts such as 'unified approach', 'self-reliance', 'ecodevelopment' and 'basic needs' tend to take on more and more dimensions and end up as fully-fledged development philosophies. This naturally irritates down-to-earth planners, trained in positivist social sciences. The theoretical polarization today is perhaps not so much between liberals and Marxists, as between positivists and normativists or between Mainstream and Counterpoint tendencies. According to the Another Development School, basic needs refers to ways of life rather than the preconditions for survival.

Manfred Max-Neef (1986), working in the Another Development tradition, makes the important distinction between needs and satisfiers, a distinction which also bridges the universal and the specific in the human needs debate. For instance housing and food are satisfiers of the need of subsistence; education is a satisfier of the need of understanding. From this point of view fundamental human needs are finite, few, and classifiable. They are also the same in all cultures. What changes (over time and through cultures) is the form or the means by which these needs are satisfied. Following Mallman (of the Bariloche Foundation) Max-Neef mentions the following needs, which according to him form a system: subsistence, protection, affection, understanding, participation, leisure, creation, identity and freedom. The satisfiers are infinite but failure to make the distinction between needs and satisfiers will only lead to a 'cosmetic improvement of the economistic view of development' (op. cit.).

5.2.3 Basic needs as development strategy

The concept of development has tended to gain in depth and richness as the ugliness and brutality of actual processes of conventional economic growth are revealed. Should this paradox be explained as some kind of escapism or is there really something to be gained from a merely conceptual development of the concept of development?

In its simpler forms, for example in the ILO declaration (at least the first of the two quotations given above), the BNA as a strategy rejects the earlier growth paradigm and argues for the incorporation of a sort of 'development guarantee' for the weakest social groups in all development programmes. The awareness of the fact that growth does not necessarily benefit the poorest is certainly a step forward. Although not a great discovery from a theoretical point of view, this is of practical importance for international development assistance and, if taken seriously, may increase its relevance.

It is interesting to note that the debate on basic needs, starting in the Third World and then subject to conservative reabsorption in the West, subsequently created suspicion in its place of origin. It is thus very common to look upon the BNA as an alleviationist strategy, a concoction by enemies of the NIEO in the industrial countries. The late Raúl Prebisch for example made this remark:

> I sometimes think – if I may be excused a touch of misgiving – that some of those who offer such formulas to the periphery from the centres do so in order to evade the problems of the new international economic order. Why listen to all this disturbing rhetoric, instead of mounting a direct attack on poverty? Would it not be easier to hand over a few funds for the purpose?
>
> (Prebisch 1980: 187)

There may of course be some grounds for such misgivings, but in principle the BNA and the NIEO are compatible and, indeed, complementary strategies. A new international economic order would be a rather immoral order without any guarantees that the transfer of resources from rich to poor countries does not stay with the rich people in the poor countries. It is also true that basic needs strategies do not make sense without such a transfer. The connection is forcefully expressed by Streeten:

> A BN program that does not build on the self-reliance and self-help of governments and countries is in danger of degenerating into a global charity program. A NIEO that is not committed to meeting basic needs is liable to transfer resources from the poor in rich countries to the rich in poor countries.
>
> (Streeten 1979: 101)

However, in the early 1980s the BNA had already lost its appeal (as the NIEO some years earlier) and in the 1981 Cancun top meeting on development, quite different (but not really new) signals were given. Now the relevance of the 'simple' BNA could be to provide arguments against a return to a development philosophy of the 1950s, a regression which is in fact favoured also by some ruling elites in the Third World. The 'complex' BNA will nevertheless remain an important component in the Another Development paradigm.

In a strong defence of the strategy as it was originally conceived by the ILO Louis Emmerij deals with the most common misunderstandings. He stresses among other things that the Basic Needs concept was conceived to be an overall social and economic development strategy, not a series of *ad hoc* projects for the poor in order to bridge temporary difficulties during a transitional period. It puts equal emphasis on growth and redistribution. It is not only concerned with the poorest and it is not opposed to the modern sector – it would rather strike a better balance between various sectors. It would reduce dependence on industrialized countries rather than increasing it (Emmerij 1988).

To conclude, it would be unfortunate if the distortions of the

BNA and the criticism of the concept that followed should remove it from the centre of the development debate as one basic dimension of Another Development.

5.3 Self-reliant development

Another key concept in the normative debate during the 1970s was Self-Reliance (SR), which, judging from this debate, meant different things to different people in different contexts.

To begin with, it connotes a hypothesized 'natural state' preceding dependence. This implies that a community enjoys SR before it gets involved in a division of labour upon which it increasingly depends for its reproduction. For most communities this is a very distant past and the concept is rarely used in this sense. Autarchy is the appropriate word.

Secondly, it may be a situation enforced upon a community when the resources it depends upon for external exchange become exhausted. Tendencies in this direction can be found in several Third World countries today. A case in point is Ghana, whose traditional exports (gold, cocoa) are diminishing while strategic imports (food and fuel) are getting increasingly out of reach for most people.

Thirdly, SR may be a way of life to which the members of a community escape when the conditions of dependence are unbearable. The runaway slaves in the Jamaica highlands (the Maroon society) would be one example. We may refer to this as internal withdrawal.

Finally, SR connotes a development strategy deliberately undertaken to promote a special kind of development. This involves the cutting of links (delinking) to the larger system of division of labour in order to avoid imposed self-reliance and find something better than autarchy. The following discussion refers to SR in the fourth sense, i.e. as a development strategy.

The popularity of this approach in the 1970s was obviously a corollary to the breakthrough of the dependence paradigm, self-reliance being the antithesis to dependence. This approach had earlier been discussed in the context of individual national experiences, such as India in the days of Mahatma Gandhi, Tanzania after the Arusha declaration and China under Mao Zedong. The concept of self-reliance was brought to the international scene by the nonaligned countries at their 1970 meeting in Lusaka and was further elaborated at their 1972 conference in Georgetown. Thus, SR emerged as a strategic concept in the international discussion just before the concept of NIEO, which provided a vision of international cooperation rather than withdrawal.

In this way two perspectives, one stressing more symmetric relations and mutual benefits from trade and cooperation, the other stressing reliance primarily on one's own resources, converged in the international debate on development in the mid-1970s. Self-reliance should, according to the NIEO philosophy, not be pushed too far, in case a more just international economic order really began to take shape. On the other hand it was something to fall back on if the rich world refused to cooperate. This more cautious attitude may be found in the Cocoyoc Declaration of 1974. A typical passage from this document reads:

> We believe that one basic strategy of development will have to be increased national self-reliance. It does not mean autarchy. It implies mutual benefits from trade and cooperation and a fairer redistribution of resources satisfying the basic needs. It does mean self-confidence, reliance primarily on one's own resources, human and natural, and the capacity of autonomous goal-setting and decision-making. It excludes dependence on outside influences and powers that can be converted into political pressure.

This formulation, which clearly emphasizes the potential value of trade and cooperation, was restricted to national self-reliance. Furthermore, it emphasizes the economic and political aspects of the strategy. As was indicated above, the approach can be widened into a fully-fledged and more radical development strategy, providing guidelines for almost all fields of action on different levels of society (Galtung, O'Brien and Preiswerk 1980).

The theoretical rationale for self-reliance as a more comprehensive development strategy has been summarized by Johan Galtung in the following hypotheses:

- priorities will change towards production for basic needs for those most in need;
- mass participation is ensured;
- local factors are utilized much better;
- creativity is stimulated;
- there will be more compatibility with local conditions;
- there will be much more diversity of development;
- there will be less alienation;
- ecological balance will be more easily attained;
- important externalities are internalized or given to neighbours at the same level;
- solidarity with others at the same level gets a solid basis;
- ability to withstand manipulation due to trade dependency increases;
- the military defence capability of the country increases;
- today's centre and periphery are brought on a more equal footing.

This list of advantages (ibid: 25–36) of self-reliance contains a certain flavour of utopianism and Galtung is at pains to point out that the thirteen rationales should be seen as hypotheses about positive effects.

One may of course also consider hypotheses about negative effects or cases of inapplicability (ibid: 17). Thus, for example, SR is a remedy against inequality only in so far as this inequality is interaction-induced. If it depends on great differences in resource endowment or if the lack of development emanates from internal exploitation, SR will not help or may even be harmful.

Self-reliance is not merely an economic policy; if consistently applied, it implies fundamental structural transformations. This becomes clear as soon as one leaves the level of rhetoric and penetrates further into the implications of the concept. Since a complete withdrawal from the international economic order is a realistic option for very few countries, our subsequent discussion will be in terms of qualifications of the concept, focusing upon the problems of size, level and degree.

5.3.1 The problem of size

As has been repeatedly stressed, the discussion on development within the tradition of development studies has been framed in terms of national development. The problem of size in economic development has been neglected in both liberal and socialist development theory. Whereas orthodox liberal theory more or less abstracts from the existence of national boundaries and presupposes a free flow of the factors of production, socialist theory is typically linked to the single experience of the Soviet Union, which can be repeated in very few countries. Obviously size has a direct bearing on the resource base for development.

The recommendations of the development strategies implied in these two orthodoxies are: either to specialize and reap the benefits from comparative advantages and economies of scale, or to develop towards national autarchy through the establishment of heavy industries, making the maximum use of the national market and domestic resources.

By gradually incorporating criticism against the unqualified version of free trade theory (originally raised by F. List), liberal theory has granted some relevance to the 'infant industry' argument. Similarly Soviet theorists have developed a theory of 'noncapitalist development', according to which reliance on the Soviet Union and a division of labour among socialist countries may be preferable to national autarchy.

The discussion on size and self-reliance has been a theoretical field of special concern for Caribbean economists, which possibly could be accounted for by geographical factors. One contribution to the analysis of the policy options open to small countries was that of W. G. Demas who saw structural transformation as the essential ingredient of self-sustained economic growth (Demas 1965). As such a transformation (implying among other things a reduction of dualism between the productivity of different sectors, elimination of subsistence production and the establishment of a national market) is hard to achieve in a small economy, Demas advocated integration of the region in a common market system. Such a policy (the creation of CARIFTA – the Caribbean Free Trade Association) has not been very successful, however, for reasons largely connected with the problem of dependence. Consequently, the attention of the Caribbean economists turned toward dependence (Girvan 1973).

It is against this intellectual background one should see the work of Clive Thomas (Thomas 1974) which will serve the purpose of illustrating a radical self-reliance approach. Thomas rejected regional integration under capitalism as a viable strategy and instead tried to develop a strategy for planned transition to socialism with special reference to small dependent economies. Here he could draw upon the Caribbean debate (although he was actually working in Tanzania when these theses were developed).

In accordance with the dependency school and its emphasis on national self-reliance, he considered the essence of underdevelopment to be the externally induced cumulative process of divergence of the pattern of resource use, domestic demand, and the basic needs of the population. Thus, the fundamental objective of a strategy for self-reliance must be to revert this process and to achieve a convergence of domestic resource use and domestic demand. To make this possible it is necessary to break with capitalism and imperialism and achieve self-reliance. A comprehensive planning of this sort, furthermore, necessitates the securing, on a political and social basis, of an entrenched worker/peasant alliance.

The converging of resource use and demand is Thomas's first iron law of transformation, the second being the convergence of needs with demand. In underdeveloped economies there is (due to both inequities of income distribution and low levels of absolute income per capita) an acute divergence between the basic needs of the population and consumption expenditure. Consumption planning should therefore aim at the 'progressive expansion of the range of socially and collectively consumed commodities' in order to reduce the distributive role of the market (ibid: 260).

Granted that this basic strategy first is implemented, there is much scope for regional integration and international trade, but it is likely that many of the small underdeveloped countries will have to advance towards socialism in relative isolation from their neighbours. The efforts at agricultural transformation and industrialization must therefore take this constraint into account.

The limited applicability of the Soviet industrialization model to small underdeveloped economies has led even radical economists to doubt the feasibility of industrialization programmes. This pessimism is, according to Thomas, unwarranted. A strategy of comprehensive planning requires the domestic production of those basic materials which are required as primary inputs for the manufacture of the basic goods of the community. It is necessary to ensure that these basic materials are substantially derived from domestic resources. This constitutes the necessary condition for the growth of an indigenously oriented technology. The scope for structural transformation will be heavily dependent on the resource configuration in each and every society. Economies of scale should be judged from the point of view of a critical minimum rather than idealized optimum levels.

As noted by Thomas, the widespread lack of confidence of the political leaders in Third World countries in the capacity of the people to master the environment, is one of the staggering consequences of underdevelopment and dependence:

> The pervasiveness of the view of the limited capacity of our people to master their environment is largely due to the society's internalization of Euro/American views of itself. As we have come to see ourselves only as others see us, so we have moved further from our freedom. (ibid: 305)

Thomas's strategy of transition can be seen as a radical version of the Listian tradition, including the economic thinking of CEPAL. He addresses the crucial question of the class base of the strategy as well as the possibility of external hostility, which makes it more realistic but also less likely to succeed. It is obvious that the whole question of self-reliance must be freed from economic nationalism in order to provide the base for a viable strategy and instead be related to different levels of economic organization.

5.3.2 The problem of level

Self-reliance must not be mixed up with autarchy. SR should rather be seen as a precondition for genuine cooperation based upon the principle of symmetry. Galtung even envisages a self-reliant world with many 'centres' but not in the Western sense of controlling a periphery dependent on it (Galtung, O'Brien and Preiswerk op. cit: 355).

This utopian image, well worth striving for, necessitates self-reliance on a number of societal levels. We may distinguish between local, national and regional levels. It is with reference to the local and regional level that we find alternative theoretical contributions, whereas national self-reliance is a mainstream political goal. An ambitious form of regionalization is the Third World collective level but due to increased interdependence and differentiation of the Third World this project now seems less realistic. The implementation of the NIEO would go a long way to achieve Third World collective self-reliance, but there is no guarantee that this would be a better world for the bulk of the population in developing countries.

Granted that this reservation is taken seriously and continuously kept in mind, there is of course nothing wrong with trade union action on the part of Third World countries in order to impose substantial concessions on the industrial countries.[74] Higher level self-reliance is, however, not sufficient for lower level self-reliance:

> Imagine regional self-reliance is implemented at the level of the Third World as a whole, at some continental level or at subregional levels. In practice this would mean full control over Third World factors, and production for and by Third World groups. But which groups? Given the present world structure the centre-periphery gradients are all there to be used by the strongest among today's poorer countries. Thus, regional self-reliance might protect the Third World against dependence on the first and second worlds (and for that reason be strongly resented by them), but would not offer any protection against penetration by the Brazils, the Irans and the Indias, and (still to come) the Nigerias. The subimperial connections of today may become the raw material for forging the imperial connections of tomorrow. (ibid: 361)

Self-reliance at the national level is, as we pointed out above, the traditional focus of interest. National self-reliance, which basically implies a strong state, is, however, perfectly consistent with a dependent or even enslaved people.[75] A certain degree of state power may nevertheless be a precondition for self-reliance on the local level: some economic planning will always be necessary even in an extremely decentralized economy, the national centre should provide a good infrastructure for any cooperation between the local units, the state may have to intervene in order to correct imbalances in resource endowments, and the local units must be protected from external penetration, for example by the multinationals.

Thus, in order to achieve self-reliance in the radically alternative sense of the word, as distinguished from the traditional preoccupation with economic and military strength of the nation state, it would be necessary to combine all three levels 'with the development of human beings everywhere as the goal' (Galtung et al 1980). This raises questions about political strategy. The answer given by Galtung is to seize opportunities as they come (i.e. when the forces favouring dependency are weak) – whether at the local, national or international levels:

> Should one then argue against the New International Economic Order which obviously represents self-reliance at the regional level only, with some excursions into the national level and with no mention of the local level, nor of the individual human beings with their needs and goals? No. It should be seen as one among many necessary but not sufficient steps, adding up to a process of fight for independence. It came about because the international level provided better opportunities for confrontation than the other two levels: with the less developed countries more conscious and better mobilized than ever before and the other side delegitimized and idea-empty.
>
> (Galtung 1976: 15)

The paths to social change are many and often mysterious. Sometimes conservative agents are as instrumental in promoting change (albeit unintended) as the radicals. The struggle for self-reliance goes on at various levels with various motives, sometimes revolutionary, sometimes reactionary. The cunning of history can only be understood in retrospect.

5.3.3 The problem of degree

As was emphasized above spokesmen for the strategy of self-reliance see this as a precondition for meaningful (i.e. symmetric) cooperation. This was implicit in our discussion on levels and size, but there is nevertheless reason to discuss the problem of degree more specifically.

It is only natural that the recent popularity of the concept of SR and the repetitive way in which the virtues of SR were praised in almost all journals in the development field (not to speak of UN conferences) provoked some doubters to come to the fore, some being in outright opposition to the ideals of SR, others being believers but feeling that a too dogmatic approach might be counterproductive.

To change an economic system from a dependent to a self-reliant one necessarily implies deep structural changes and, as once was emphasized by C. Wright Mills, there is always a connection between structural and individual adjustments (Mills 1959). It is therefore important that the individual fully understands what is going on. Hence the emphasis on participation and redistribution in strategies of SR (Rothstein 1976). This will serve as a check on too dramatic crash programmes for self-reliance which, if imposed on the people, tend to backlash on the government.

As also was emphasized in the Dag Hammarskjöld report referred to above, a strategy of self-reliance will necessitate selective participation in the international system:

> Selective participation of Third World countries in the international system is a prerequisite for the application of new development strategies, for strengthening internal sovereignty and for reinforcing self-reliance.
>
> (What Now: 67)

It is further underlined in the report that the concept of selective participation is a flexible one. Different conditions among Third World countries will require a variety of selections not only by different countries but also by any country at different times. The only general guidelines that can be formulated are the following ones:

- There is a minimum degree of links required to sustain the development process.
- There is a maximum degree of links beyond which no effective sovereignty can be maintained.
- There are affirmative links which reinforce self-reliance.
- There are regressive links which weaken self-reliance.

It is an important task of alternative development theory to provide the necessary theoretical and empirical base for the strategy of selective participation with the purpose to promote self-reliance on all levels. This will require knowledge of the development process in individual countries and about the international context, a theory about which links are productive and which are counterproductive, and futuristic studies of the possible outcome of various strategies which obviously will be undertaken by the centre. As the centre should not be conceived of as a monolithic block, there is also a need for research about cooperation and conflict within the centre and its implications for self-reliant strategies.

5.4 Ecodevelopment

One fundamental incentive behind the current search for another development paradigm is the new environmental consciousness that emerged in the 1970s and got its most recent expression in the report of the World Commission on Environment and Development: 'Our Common Future' (1987). Consequently this ecological orientation constitutes an important dimension in Another Development: ecodevelopment or sustainable development (Redclift 1987).

This section first discusses the recent preoccupation with resource scarcity in the social sciences, a concern which fits badly with the orthodox paradigm. Secondly the question is raised how well-prepared the social sciences are for dealing with ecological issues. Thirdly the relationship between development and environment is discussed.

5.4.1 The ghost of scarcity returns

Modern growth and modernization theory has been relatively unconcerned with the problem of scarcity. This optimistic streak is different from the classics for whom the problem of scarcity was a major preoccupation in political economy, not only in the Ricardian (relative) but also in the Malthusian (absolute) sense. That most of the pessimistic prophecies of the 'dismal science' fortunately turned out to be wrong, or at least exaggerated, allowed later generations of economists (the exception of Jevons and the 'coal question' proves the rule) to be much more relaxed on this issue. The warnings have come from other quarters, whereas the economists, typically, have been trying to keep up the optimistic spirit (Beckerman 1974).

In Mainstream Marxist theory scarcity is socially determined by the shackles imposed on the forces of production by the capitalist system, whereas communism is characterized by abundance which removes conflict over resource allocation and eliminates the role of economics (Nove 1983). To point out this bias in Marxism, which by the way is subject to some rethinking in the era of 'new scarcity', is not to deny that scarcity takes a particular and probably more severe form under capitalism where misery and affluence coexist in an outrageous fashion.

We are told (Parsons 1977: 67) that in a transition from a capitalist to a socialist ecology the mastery of nature should:

- benefit all people and not just a small ruling class;
- maintain the dialectical balance of natural ecology in harmony with human needs;
- be qualified by a theoretical understanding and an aesthetic appreciation of nature.

This kind of transition is yet to be seen.

However, absolute scarcity does exist and this necessarily implies 'limits to growth' (Brookfield 1975). The problem is what limits to what kind of growth, and in what time perspective. These elusive questions may have been unsatisfactorily answered but that is certainly no reason to deny, suppress, or neglect the issue of scarcity.

Apart from physical limits to growth we are also reminded that there are social limits to growth:

> Where the social environment has a restricted capacity for extending use without quality deterioration, it imposes social limits to growth.
>
> (Hirsch 1976)

The polarization between doomsday writers and incorrigible growth optimists gave the problem of scarcity its ghostly character in the 1970s. It is now well-documented in statistical reports but yet it somehow remains unreal. The 'automobilism', for example, which more than anything else illustrates the arrogance of industrial man and his ignorance of the problem of scarcity, still dominates the industrial world and develops at a very high speed in the rest of the world as well. At the same time the warning signs continue to reach us in a steady stream: polluted air, poisoned lakes and seas, dying forests, etc.

The first alarm report on the physical limits to growth was the 1972 report from the Club of Rome which then set the framework for the discussion in the 1970s, much of which happened to be highly critical of the report. Perhaps the obsession with physical limits, implying a sudden apocalyptic end of the earth, was its main weakness. The fact of relative scarcity will of course make itself felt in a number of ways long before a society approaches any absolute limits. Some of the resultant changes may relieve the scarcity problem through technological innovations, others will aggravate it through intensified competition and political/military struggle on various levels of the world system.

Recently the relationship between environmental degradation and conflict has drawn the attention of researchers and activists. For instance an Earth-scan report (Timberlake and Tinker 1984) argues that diminishing natural resources have become an important cause of violent human conflicts both between states and within states (riots, military coups and revolutions).

The relationship as such is obvious since the changing environment provides the context in which conflicts emerge, but the actual cause–effect relationship has not been the subject of systematic research. The problem of environmental refugees in Africa, the Indian subcontinent, Central America and the Caribbean, and the ungovernable cities in Third World countries are cases in point, showing the undeniable connection between scarcity and conflict.

The conflict in Central America has, above the superpower involvement, large elements of environmental causes, such as agricultural labour underutilization in El Salvador, coupled with deforestation, soil erosion and diminishing subsoil water resources. Another aspect of the same problematique could be found in rural unrest which has its origins, among other things, in migration from degraded land to better land. This type of migration can and in certain cases will be mixed with political and religious issues, as in the case of Punjab or at the borders of Ethiopia, Kenya and Sudan. The same problem can be found in competition concerning forests. The competition between a larger area of arable land and fuel wood, or between commercial forestry and social forestry in India, Brazil and elsewhere becomes acute when the resources are scarce.

Environmental degradation and diminishing shared renewable natural resources are an important cause of violent conflicts and forms a crucial part of the poverty complex. These causal factors have been neglected in the scientific community, particularly in so far as the political economy of environmental degradation is concerned (Blaikie 1985).

5.4.2 Ecology and the social sciences

Theories about the relationship between ecology and development, as well as discussions about environmental problems in the context of development strategies show important concerns that only recently have been generally acknowledged. It still remains to be seen how these theoretical efforts will affect development thinking and praxis.

So far the experiences suggest that the marriage between the two does not come about easily, in spite of many programmatic declarations at conferences and workshops. One wonders why – is there some hidden contradiction or incompatibility between, for example, economic and ecological thinking? Kenneth Boulding, who has been reflecting a lot on both economy and ecology, identifies several similarities in their methodological approaches, but also significant differences:

> In their underlying value systems, ecologists often tend more toward Hindu or Buddhist values, in which man is seen as only part and perhaps not even a very important part of a vast natural order. Economics, however, emerged out of a civilization, part of Western Europe, that was created largely by Christianity and which regarded man as the measure of all things and the universe as existing mainly for his pleasure and salvation. (Boulding 1966: 230)

The distinction in terms of values goes much deeper than a contrast between the economic and ecological disciplines. It corresponds to that dualism in Western thought we earlier referred to as 'Mainstream' v. 'Counterpoint'. Economics is certainly the most mainstream of the social sciences, although it is true that the controversy as to whether or not natural resource scarcity and economic growth are fundamentally antagonistic is a classical one. The dominant line of thought takes an optimistic view of technological advance and regards natural resources as a highly dynamic concept. The ecological movement is consequently placed within the realm of 'metaphysical naturalism' (Barnett and Morse 1963). In the late 1960s however, there was an animated debate partly stimulated by E. J. Mishan's The costs of economic growth (Mishan 1967, Weintraub et al 1973).[76]

As far as the possibility of future growth is concerned we have to distinguish between scarcity of natural resources in the physical sense and the ecological or environmental functions that different ecosystems provide (Hueting 1980). To understand when such functions are threatened economists need help from the ecologists.

Until recently the problem of interactions between ecology and society has been conspicuous by its absence from the other social sciences as well – with the exception of social anthropology, where the man – environment relationship and its ramifications have been more difficult to hide than in social sciences which mainly deal with industrial, functionally organized societies and macro-structures. It is well known that many anthropologists, generally in vain, have emphasized the local interests of populations in connection with dams or other large-scale constructions. 'Progress' most often had its way, however.

Due to this neglect of the natural base of human societies, only widespread fear of a global ecological crisis could bring social scientists and others dealing with development problems to analyse development as a process involving both society and the environment. Hopefully, the present upsurge of interest in ecological aspects of development is not merely a passing intellectual vogue.

During the 1970s rethinking started in several social sciences. Political scientists like Karl Deutsch began to speak about ecopolitics: the political challenge produced by the growing interplay of man's economic activities and the environment (Deutsch 1977). This challenge is not only the concern of nation states. Environmental interdependence transcends political borders, thereby increasing the importance of international actors and institutions.

Thus ecology also provides a new context for international relations, what Dennis Pirages has called 'global ecopolitics' (Pirages 1978). The new scarcity, or the problem of planetary finality, poses new issues in world politics. These issues underline the basic oneness of the world from an ecological point of view. This also implies a normative bent in international relations theory: The 'is' of power politics is being replaced by the 'ought' of ecopolitics. Furthermore, there are now a vast range of actors apart from the states that create the substance of world politics which tends to be so much more than the 'high politics' of national security.

One should expect spatially and resource-oriented disciplines like human geography to shoulder a particular responsibility in forging the ecosocial system and the issue of development into an analytical whole. Harold Brookfield, for example, has observed that developments studies and the study of changes in man's use of environment together provide theoretical elements to be used in the task of generating a dynamic man/environment paradigm (Brookfield 1975).

The new discipline of human ecology, now emerging as a

transdisciplinary field of training and research, could possibly contribute to such a paradigm, provided it is firmly established as a combined study of man living in two systems, the ecological and the social, held together by a high degree of interdependence.

So far, development strategies have been ecologically blind, but ecology will certainly be incorporated in any future development paradigm, so as to permit a systematic analysis of environmental aspects of social and economic change. When this is done, it will become evident that many contradictions are involved, which are now hidden behind the universal but rather superficial agreement that ecological catastrophes are unpleasant.

In terms of regional planning ecodevelopment or sustainable development is a process of development that is more 'territorial' and less 'functional'. This distinction refers to two different paradigms in regional planning which stand out most clearly in the US tradition (Friedman and Weaver 1979).

> When we review the changing course of regional planning doctrines over the past half-century, two major forces of social integration appear to alternate with each other: territorial and functional. Intertwined and complementary to each other, they are nonetheless in constant struggle. The territorial force derives from common bonds of social order forged by history within a given place. Functional ties are based on mutual self-interest. Given inequalities at the start, a functional order is always hierarchical, accumulating power at the top. Territorial relationships, on the other hand, though they will also be characterized by inequalities of power, are tempered by the mutual rights and obligations which the members of a territorial group claim from each other.
> (ibid: 17)

The territorial and functional forms of social integration (which can be compared to more famous distinctions of types of social organization such as 'Gemeinschaft' v. 'Gesellschaft') thus both complement and contradict each other. A complete dominance of the territorial principle would spell the end of the nation state but this is a prospect that we probably do not have to worry about. The problem is rather how to decrease the dominance of function over territory, and how a 'recovery of territorial life' can be brought about. To some extent this already takes place when people react in their own way and as best as they can to the challenges of the crisis of the world economy taking the local space as their point of departure (Stöhr 1988). The point however is to make the territorial principle part of a planned transition towards a more sustainable development.[77]

Sustainability is a principle that has appeared in development theory as a consequence of the environmental concerns from the early 1970s onwards. It is particularly associated with Lester Brown and the Worldwatch Institute, and the main message it carries is that neither the old nor any new international economic order would be viable unless the natural biological systems that underpin the global economy are preserved (Brown 1981). The starvation crisis of Africa is a grim reminder of this fact. Deserts are spreading not only in Africa but in the Middle East, Iran, Afghanistan and India.

Lester Brown has underlined four areas which are problematic from the point of view of sustainability:

- the lagging energy transition;
- the deterioration of major biological systems;
- the threat of climate modification;
- global food insecurity.

Although the depressed world economy has made the development of energy demand less dramatic than projected, the transition to new sources of energy is still a necessity. Both nuclear energy and coal create sustainability problems of their own, and the development of renewable energy sources has proved to be a slow process.

The deterioration of the four major biological systems – oceanic fisheries, grasslands, forests and croplands – is another serious problem, since their 'carrying capacities' are exceeded. Thus, not only nonrenewable but also renewable resource bases are shrinking.

Climate changes occur both as long-term cyclical processes and as a result of human activities (deforestation, pollution of the atmosphere, etc.). The effects of even minor decreases or increases in temperature may have catastrophic effects upon the productivity of various ecosystems.

A transfer of cropland from subsistence production to cash-crop production and a shift from a variety of indigenous food crops to imported foods like wheat and rice has created a global food insecurity. These ecological threats to human security are clearly related to the Mainstream paradigm of development and can only be countered if this paradigm of development is reconsidered.

According to the dominant functional principle, development is an abstract process related to an artificial 'national economy', simply an aggregate of production data. Behind this abstraction we can observe the concrete socioeconomic 'worlds' that most people identify with and depend upon. When this relationship is disturbed and threatened by 'modernization', conflicts occur over the goals and means of development. Smaller territorial units, however, lack autonomous power in the functional system, typically organized as a centre-periphery structure. The revitalization of territorial life would therefore not be possible without a transfer of power to local communities, while the state level assumes the role of a coordinator.

It is also necessary to strengthen territorial units larger than nation states and, ultimately, decision-making at the global level as the only way to protect the global ecosystem and achieve sustainability. Nation states that specialize and become efficient in exploiting certain ecological niches tend to continue this activity regardless of the signs of danger, an unfortunate result of that particular blindness which is associated with the promotion of growth in a functional system with little regard to what happens to specific ecosystems.

5.4.3 Between growthmania and ecologism

The ghost of scarcity has come to stay. The problem of development in the context of ecological constraints is therefore something the social sciences will have to deal with in spite of their lack of preparedness. The orthodox view of unlimited economic growth as some kind of natural law must be replaced with a sense of historical relativism. One must understand that development is a concrete social process with specific causes and preconditions which are liable to change due to the relative power balance between different 'social projects'.

On the other hand, a general hostility towards growth because it threatens the ecological balance must also be avoided, since there is no point in substituting one myth for another. The road forward consequently goes somewhere between growthmania and ecologism. This awareness of threats coming from both Scylla and Charybdis is very explicit in the theorizing around the concept of ecodevelopment coming from Ignacy Sachs.

The notion of ecodevelopment was born at the UN Conference on the environment in 1972. The innovator then was Maurice Strong, and later the concept was developed and popularized by Ignacy Sachs, who has suggested this definition:

> . . . ecodevelopment is a style of development that, in each ecoregion, calls for specific solutions to the particular problems of the region in the light of cultural as well as ecological data and long-term as well as immediate needs. Accordingly, it operates with criteria of progress that are related to each particular case, and adaptation to the environment plays an important role.
>
> (Sachs 1974: 9)

In many respects this approach constitutes a far more radical challenge to the modernization paradigm than does the dependency perspective. Not only positions in a hierarchy are questioned, but the very values determining the hierarchical order. There are, according to this view, no models to emulate. A 'backward' country should not look for the image of its own future in the 'advanced' country but in its own ecology and culture. Development has no universal meaning. There is no development as such, only development of something, which in this case would be a certain ecoregion. A development strategy informed by this perspective must make efficient use of those resources which happen to exist in that particular area, in a way that both sustains the ecological system (outer limit) and provides the people living there with their basic human needs (inner limit). Efficient use, however, also includes exchange, as long as the principle of sustainability is complied with.

Sachs himself mentions three pioneers behind this emerging tradition: Benjamin Franklin, Mahatma Gandhi and René Dubos (Sachs 1980: 16–19). Franklin represents innovation, the attitude of making the most efficient use of resources existing in the surroundings. Gandhi symbolizes the ethical imperative in development, the principle that development first of all must improve the conditions of the poorest. To René Dubos the most important thing is the symbiosis between man and nature, a symbiosis that avoids both growthmania and ecologism. To disturb the ecological balance (a crime according to ecologism) is not necessarily harmful since such disturbances form part of ecological change, which is and always has been a perfectly natural process. However, certain ecological changes may lower the productivity of the ecological system in a significant way and do irreparable damage to important ecological functions. There should be no disagreement about avoiding such changes. The point is simply that there must be more awareness of the ecological implications of alternative development strategies before the harm is done.

The paradigmatic quality of the ecodevelopment approach is shown in the way it subsumes other elements of alternative development thinking, or Another Development. We have already referred to the basic needs approach implied in Gandhian ethics. It is rather obvious that self-reliance (in its more radical emphasis on local self-reliance) constitutes another fundamental principle in the ecodevelopment strategy. It also shares with other elements of Another Development a rather extreme normativism. Ecodevelopment does not really take place in many parts of the world today, but according to its spokesmen the world would be a better place to live in if it did. The problem, of course, is how this can be achieved. As Bernhard Glaeser and Vinod Vyasulu rightly point out, the concept is utopian since it does not tell us how to get there:

> Ecodevelopment projects must include the power variable from the very beginning. Who has power to mobilize resources, to corner benefits, to stall the process? (Glaeser 1984: 35)

This observation is pertinent since a typical ecodevelopment project does not add to the mobile resources of the state and would thus not be in the immediate interest of the ruling elite.

Ecodevelopment necessitates a development strategy which differs radically from conventional strategies with their universal elements: capital, labour, investment, etc. An ecodevelopment strategy, in contrast, consists of specific elements: a certain group of people, with certain cultural values, living in a certain region with a certain set of natural resources. The goal of an ecodevelopment strategy, then, is to improve that specific situation, not to bring about 'development' in terms of GNP or some other abstraction.

Ecodevelopment can be applied to a great variety of local ecosystems and it is only by so doing that the abstract quality of the concept is replaced by a concrete understanding of its implications. Let us therefore relate our concluding discussion to the forest sector. Forestry clearly illustrates the contradiction between First and Second System interests on the one hand, and Third System interests on the other.

Typically the rise of state power in areas commonly designated as underdeveloped – a process started in the colonial era – made natural resources such as forests a 'national' asset. In practice this meant an encroachment by the First and Second Systems upon local rights which were historically given rather than being juridical conventions. Forest departments guaranteed railway companies, timber contractors, paper mills, etc. access to forest wealth, while playing a police function *vis-à-vis* the local communities, which traditionally had been using forests for satisfying a great variety of needs: shelter, food, grazing, fuel, small timber, ornaments, rituals, recreation, etc.

The local community had its indigenous way of protecting the forest through social and religious constraints, although it did not always succeed in periods of population growth or ecological changes. The protective role of forest departments, however, has been far less conspicuous than its exploitative role. The results can be seen today. In India for example, the forest area is reduced to twelve percent of the total area. It is difficult to see how the increasing demands shall be satisfied from this shrinking area, in which there already are legitimate interests, such as forest tribes and village communities.

This situation, which is rather typical in the Third World, obviously calls for a new approach and there is consequently no lack of recipes: community forestry, farm forestry, social forestry, etc. These concepts reflect the need for an ecodevelopment approach but the confusing way in which they are used and applied in concrete situations reveals a poor understanding of what such an approach really implies.

Ecodevelopment means that the local community and the local ecosystem develop together towards a higher productivity and a higher degree of needs-satisfaction, but above all that this development is sustainable in both ecological and social terms. A program of tree planting that mainly serves commercial needs, that is carried out on rich farmers' private lands, that leads to a forest monoculture (e.g. eucalyptus plantations), and that negatively affects both agricultural production and the level of employment is contrary to ecodevelopment by contributing to environmental deterioration and underdevelopment.

However, even well-intentioned social forestry programmes have tended to go in that negative direction for the simple reason that community, if it exists at all, is seriously undermined by the development of an inegalitarian class structure, ethnic conflicts, and encroachment by external interests. Therefore there are substantial problems involved in developing a new kind of participatory forestry. Participation and mobilization, however, are not ends in themselves. They are a necessary means if a replantation programme that serves local needs rather than market demands shall succeed.

Market demands are certainly not illegitimate – pulp and paper are essential for cultural and educational improvement – but they must be balanced against local needs, and in fact the latter must have priority if we are talking about ecodevelopment, and even more so if we link this with egalitarian development. It is rather obvious that a community forestry programme needs a functioning community and that it succeeds only to the extent that the community is viable. This would still hold true for many forest tribes.

Greatly stratified villages and urban or semiurban areas, on the other hand, would need action on some kind of class basis, preferably aided by external but still territorially based networks of activists and voluntary development workers – the Third System in action. Third System politics is always territorially oriented and so is ecodevelopment.

5.5 Ethnodevelopment

A development strategy that – as Julius Nyerere once proposed – concerns 'people not things' will for all practical purposes have to deal with ethnic groups, since 'people' consist neither of individuals, nor of nation states. One of the unexpected and still not very well understood outcomes of the mainstream development process is an explosion of ethnic violence. The way ethnic conflicts or ethnic movements relate to development is highly complex, but the differential outcome of both growth and stagnation must obviously have an impact.

In what follows we shall first deal with some recent incidents of ethnic disturbances, secondly discuss how these cases may relate to various patterns of development, and finally explore the preconditions for a model of 'ethnodevelopment', i.e. principles of development that bring out the potential of different ethnic groups rather than bringing them into feuds.

5.5.1 The rise of ethnopolitics

Most people belong to an ethnic group, and should this not be the case ethnicity can, if necessary, easily be manipulated and even invented for political purposes.

By ethnopolitics I mean ethnic identity activated and used for the purpose of political mobilization. As a phenomenon politicized ethnicity is not new. What is new is the sudden wave of ethnic tensions and ethnic movements that we now witness in the three 'worlds'.

> From the Canadian glaciers across to the moors of Britain and to the high Pyrenees, down through the broad belt of Black Africa, the arid hills of Turkey, Iraq, and the southern USSR to the riverbanks and deltas of South Asia, peoples are challenging nationstates. In one of the most perplexing trends of the second half of the 20th Century governments are being hounded, cajoled, and defied by minorities within their societies – by ethnonationalism.
>
> (Shiels, 1984)

Ethnic conflicts are worldwide. They are endemic in Africa but have been particularly intense in South Asia during the 1980s. Pakistan was divided once on the ethnic issue. Today Karachi is often paralysed due to ethnic wars between Pathans and Mohajirs, and on the national level the Punjabi dominance is challenged from Sind and Baluchistan. Bangladesh itself today experiences ethnic conflicts between Buddhist Chakmas from the hilltracts and Muslim Bengalis from the overpopulated plains. History ironically repeats itself in the form of new patterns of internal colonialism. Similar conflicts take place in India between tribals and the Hindu population, for instance in Bihar. In Assam the 'sons of the soil' defend their primary right (as they conceive it) to resources and employment against immigrants from Bangladesh, West Bengal and Bihar. 'Sons of the soil' organizations are emerging in many states, for instance Maharashtra and Karnataka. In Gujarat the higher castes attack lower castes because of the policy of protective discrimination. In Kashmir and Andhra Pradesh, Muslims and Hindus kill each other; in Punjab, Hindus and Sikhs; and in Sri Lanka, the Sinhalese-Tamil conflict for years bordered on civil war.

In all these conflicts there is an economic factor, of varying importance but never absent. However, there is no uniform economic cause behind. For the forest tribes in different parts of India their mobilization is a matter of physical survival. For the Sikhs in Punjab the economic interests involved are harder to specify and differ according to class interests. The economic relations between the Sinhalese and the Tamils in Sri Lanka are also complex. From the perspective of class it is necessary to distinguish between several subgroups within the two communities.

However, the reality of ethnic conflicts can no longer be disregarded by theorists and practitioners of development, since the rise of ethnopolitics must be related to economic development one way or the other. The problem is that we do not really know how. I therefore agree with Stavenhagen when he says that the neglect of the ethnic question in development thinking is not an oversight but a paradigmatic blind spot (Stavenhagen 1986: 77). In order to build a theory of Another Development one must explore the ethnic factor in development within a rather simple framework to start with. The time for generalizing and theorizing will come later.

The main divergence from conventional development theory is the assumption that people are divided in territorial cultural groups as well as being individual consumers and producers, buyers and sellers, employees and employers, etc., and the strong emphasis given to the former type of identity rather than the class-based.

5.5.2 Ethnic relations and development

In what ways can conflicts between ethnic groups and between the state and specific cultural groups be causally related to different patterns of economic development? The range of economic problems that may influence ethnic relations is great indeed: struggle for scarce resources, regional imbalances, infrastructural investments with a great impact on local economic systems, exploration of 'new frontiers', labour market conflicts, distributional conflicts, etc., etc.

Most of these problems affect all societies but in multiethnic societies they are more severe and tend to become permanent. There may be spread effects within ethnic groups, less so between them. In order to provide some order one could work with the following tentative distinctions:

- conflicts over the control of natural resources;
- conflicts relating to major infrastructural projects affecting local ecological systems;
- conflicts stemming from impersonal and secular but uneven trends such as commodification, proletarianization, urbanization, etc.;

- conflicts concerning the principal content of national development strategies;
- conflicts of state-controlled redistribution.

Conflicts over natural resources can be exemplified by the way forest wealth is used by jungle tribes on the one hand, and urban middle class populations on the other. For the former the forest represents a way of life, for the latter it means for instance building materials or paper for the newspaper industry. Urban populations may not even be aware that a forest is inhabited by human beings, and if they were they may not consider these beings as human. The conflicts resulting from such clashes of interests and world-views are fundamental. They represent two different paradigms of development: growth and modernization v. ethnodevelopment or, in different terms, functional v. territorial development.

The same applies to the second type of development-related conflict: infrastructural projects affecting local ecosystems. Such projects, although carried out in the name of development, may be ethnocidal in their consequences.

The third type of conflict has to do with the unevenness of development which means that certain regions are placed in more advantageous positions than others and consequently attract more investment and skills. Such centres usually become the bases for nation-building, whereas people in the backwash regions are the reluctant citizens. Their protests are normally expressed in ethnic terms. The state tends to develop common interests with the most commercialized regions since they provide the type of free-floating resources upon which various state functions depend. Furthermore the 'state class' is often recruited from the same ethnic group, which reinforces the bias.

Most of the options we have referred to so far are more or less explicitly reflected in the development strategy carried out by a particular regime, but in some cases the ethnic issue is part and parcel of the development strategy and therefore directly intervening in the ethnic struggle. This is for instance the case with the bhumiputra policy of the Malaysian government with the purpose of strengthening the economic position of the Malays *vis-à-vis* the Chinese.

Finally government directly intervenes in the ethnic struggle by the way it redistributes resources: public works, education, employment and patronage. As is the case with the policy of protective discrimination by the government of India the allocation of education and public employment opportunities by quotas even makes it necessary to belong to a certain community and remain a loyal member of it. In Africa state coherence has been maintained through distributive economic policies (Hydén 1984). Thus, the economic crisis is also a crisis for the state, implying ethnic conflicts.

5.5.3 Development and cultural pluralism

Development theorists have always stressed the importance of 'noneconomic' factors, only to forget about them after having made the obligatory reference. The reasons are easy to see as soon as one tries to define culture. A culture is the fundamental condition for collective existence, it is the unconscious universal frame of reference, which becomes specific only in confrontation with other cultures. This in turn starts a process of objectification and reification which changes both the content and function of culture.[78] From being to its bearer what water is to fish culture becomes a weapon of defence, a source for political mobilization, and sometimes, unfortunately, for arrogant ethnocentrism. Upon deeper probing 'cultural development' thus turns out to be an ambiguous phenomenon.

In this context our often-referred-to distinction between functional and territorial principles of development is of help. According to the former, development is basically a result of specialization and an advanced division of labour between regions; according to the latter, it is the regions themselves which are to be developed, not the larger functional system. The territorial approach is inherently cultural.

It is easy to see the connection between these two principles on the one hand, and ethnic conflicts on the other. Ethnic groups are most commonly locally based and their cultural identity is closely related to the ecological particularities of the region and to a certain mode of exploiting the natural resources. A process of 'development' that threatens the ecological system of a region is therefore also a cultural threat against the ethnic group for which this region is the habitat. Obviously such a process cannot be regarded as development for the ethnic group thus threatened, even if this were considered to be development in the macrosystem, or functional system. As Rodolfo Stavenhagen has suggested, a development process appropriate for a particular ethnic group can be called 'ethnodevelopment' (Stavenhagen 1986). This is a most radical concept since it turns the tables on the conventional conception of ethnicity as an obstacle to modernization (Thompson and Ronen 1986: 1). There is in most social science literature an inbuilt bias against ethnic identification and in favour of national identification, regardless of how unrealistic a particular nation-state project may be. But the re-emergence of ethnicity may also be seen as a 'reaffirmation of a long-existing ethnic identity in the process of positive development – as an integral part of development, where the state (or at least certain aspects of it), not ethnicity, is an obstacle to development' (ibid: 6).

I consider ethnodevelopment as a basic component in Another Development, together with egalitarian development, self-reliant development and ecodevelopment, or sustainable development. They are all complementary and mutually supportive.

Egalitarian development implies development consistent with basic needs and self-respect; self-reliant development is based on the principle of autonomy; and ecodevelopment refers to the habitat where ethnic groups live. Thus ecodevelopment and ethnodevelopment are two aspects of the same thing in situations where ethnic identity is territorially based.

In the case of dispersed, i.e. nonterritorial, ethnic groups this does not apply, and the principle of ethnodevelopment would rather be expressed in the protection of cultural, religious, and linguistic rights, in the framework of a functional system. The concept of culture is, however, basic to the two situations (the territorial and the functional).

Sometimes it is asserted that a development strategy must take the culture of a specific country into consideration. The 'national' culture is however often rather artificial compared to regional and ethnic cultures, unless one particular subnational culture is elevated as the national culture. Thus, in any case, ethnodevelopment is a challenge to the nation state. It is development within a framework of cultural pluralism, based on the premise that different communities in the *same* society have distinctive codes of behaviour and different value systems (Worsley 1984). This conception of culture could be contrasted with a hegemonic concept of culture as diffused downwards and assumed to result in a shared national culture, but in reality implying ethnocide.

This is the fundamental contradiction in the nation-state project, a contradiction which cannot be resolved unless this project is redefined. Otherwise the state must be described as an agent of anti-development. The contents of a development strategy enforcing cultural variety and ethnodevelopment is easily spelled out: decentralization, participation, rural rather than urban bias, territoriality, self-reliance, ecological balance, etc. The ways states are formed and ruled are, however, generally not conducive to cultural pluralism. This is the underlying cause of the ethnic violence referred to at the beginning of this section. Unpleasant as it may seem the current wave of ethnic violence may have the historical function of modifying the nation-state project and the pattern of development inherent in it.

Chapter 6

Transcending the European model

> Europe is the continent that gave birth to the nation state, that was the first to suffer its destructive effects upon all sense of community and balance between men and nature. It is the continent which, therefore, has every reason to be first to produce the antibodies to the virus it itself generated.
>
> DENIS DE ROUGEMONT

The theme of this book is that development theory grew out of a gap between, on the one hand, evolutionistic growth and modernization theories, derived from the Western experience, and, on the other, the persistent underdevelopment in most of the Third World. The methodological, conceptual and theoretical filling of this gap in my view constitutes what the development of development thinking fundamentally has been concerned with, and should consequently be seen as the main source of theoretical enrichment.

This chapter concludes our survey of the development of development theory by bringing us back where we started, in the 'developed' world. It is a world that nowadays is less confident, less prone to hold up itself as a model for others, and slightly worried about its own future. This goes for both East and West. We have noted that many Third World thinkers assert that there is nothing to imitate any more. This is a healthy reaction. But can the so far dominant model be transcended by its own creators – or, rather, those who have inherited it?

6.1 Development theory returns to Europe

An influential early criticism of development theory, discussed in Chapter 2, was an article by the late Dudley Seers from 1963. The title of this now classic piece was 'The Limitations of the Special Case'. It raised the awkward question whether conventional economic theory was valid for Western industrial capitalism only.

This question gave a generation of development economists something to reflect upon.

Two decades later, when this position had become part of conventional wisdom, Seers went further with the argument that neoclassical theory, as well as crude Marxist theory, had a limited applicability in the First World as well.[79] His point was that the insights achieved in development studies through its manifold experiences of different structural situations and varying development problems could be of relevance for the developed countries as well.

One example is the case of the European periphery, since there are obvious structural similarities between the fringe areas of Europe and the global periphery. We shall come back to this. But why not take a further step? Are there perhaps other issues, analysed by development theorists normally concerned with the Third World, that have a bearing upon contemporary problems in the industrialized countries? Is self-reliance a good strategy for Sweden? To what extent are basic human needs fulfilled in Denmark? Could Poland have an alternative development? Should the Soviet Union adopt an 'agriculture first' strategy? Can development in Germany be more in tune with the ideals and principles of ecodevelopment? What about the future of modernity in Europe?

6.1.1 The future of modernity

The shortcomings of modernization theory in analysing development problems and promoting development in the Third World were extensively discussed in Chapter 2. Since the main weakness of modernization theory was held to be its Eurocentric bias, one would expect it to be more relevant to Western development and the future of Europe. This is not the case. In modernization theory the future of modernity is left completely blank: 'the advanced countries of the West, it was assumed, had "arrived".' (Huntington 1971: 292)

From the early 1970s the prospects of Europe began to look less bright. The optimism of the modernization paradigm was replaced by equally unrealistic doomsday prophecies and, as summarized in Chapter 1, a renewed interest in the phenomenon of 'crisis'. Some stressed the need to 'correct development theory's neglect of the future of modernity' (Lindberg 1976: 4).

This brings up the connections as also the distinctions between development theory and futures studies. Development theory is concerned with a not too distant future, continuously shaped by human actions of today and yesterday, and analysed in the light of certain more or less explicit normative criteria. It is less concerned with long-term projections of a few specific variables, and more with the short-run interplay between a larger number of variables. Development theory thus deals with the implications for the future of current processes, and the more or less unconscious choice between different futures inherent in alternative courses of human action.

If future studies, depending on the assumptions, analyse how far from now the world will reach fulfilment or come to an end – if things go on as before – development theory is more concerned with why things go on the way they do, and what could be done about it. Man is a problem-solving animal, and development theory is fundamentally about resolving institutional and structural problems.

This rough comparison is perhaps a caricature of futures studies which, it should be remembered, is a rather new discipline developing in different directions. Galtung, for example, expected a growing similarity between futures studies, peace studies and development studies: 'peace, development and the future are all increasingly seen in terms of human self-realization' (Galtung 1977: 6). In retrospect it seems more correct to think of these disciplines as themselves split between different paradigms with some subtraditions converging.

The new art of global modelling is a good example of the impact of paradigm change. As noted earlier, the most discussed world models in the 1970s were 'The Limits to Growth', 'Strategy for Survival', 'The Alternative World Model' (Bariloche Model), and 'The UN World Model'. In view of what was said above it comes as no surprise that the assumptions and results of these different efforts at global modelling were contrasting in fundamental respects. The various models were based on different (usually implicit) theories, some presumed to be positive, others explicitly normative.

It is the light that these models throw on the underlying theories and values, rather than the technicalities of them, that is of importance in this context. Of course the mathematical analysis of how a set of variables affect each other, and the system they define, may appear as an objective exercise simply because of the use of the computer. However, the choice of variables is clearly a normative problem. It is for example not without importance if population growth in the South rather than overconsumption in the North is used as the strategic variable. The Bariloche Model was explicitly normative, and consequently closer to normative development theory than the other models.

Even the concept of 'postindustrial society' can be understood in two basically different senses and it has been suggested that these two interpretations may be the poles of the ideological debate in forthcoming decades (Marien 1977: 461). The two visions correspond basically to what I have described as Mainstream v. Counterpoint.[80]

The Mainstream variety is best exemplified by Daniel Bell's classic work on postindustrialism (Bell 1973) and, in extreme form, in the works by Herman Kahn, as is evident from the following quotation:

> At the midway mark in the 400-year period, we have just seen in the most advanced countries the initial emergence of superindustrial economies (where enterprises are extraordinarily large, encompassing and pervasive forces in both the physical and societal environments), to be followed soon by postindustrial economies (where the task of producing the necessities of life has become trivially easy because of technological advancement and economic development). We expect that almost all countries will develop the characteristic of super- and postindustrial societies. (Kahn 1976: 1)

A rather different interpretation of postindustrialism refers to a future society emerging after the expected breakdown of industrialism – or the voluntary dismantling of it. This line of thought sometimes relates to the pessimistic view of the future of modernity implied in Max Weber's iron cage of bureaucratization and the hegemony of 'Zweckrationalität'. This classical debate is revived in Habermas's work on modernity or 'the modern project'. We have already referred to his analysis of the contradictions of the welfare state in Chapter One. For Habermas the modern project above all corresponds to the heritage of Western rationalism, which he actually wants to defend against contemporary philosophical tendencies towards relativism (Bernstein 1985, particularly Giddens's contribution). Thus Habermas tries to distinguish the modern project as such from its present pathologies. This is in contrast with more fundamentalist critics of the modern project, for instance Rudolf Bahro who identifies it with 'exterminism' (Bahro 1984).

The Ecologist magazine in 1974 took as its subtitle Journal of the Post-Industrial Society. This kind of 'postindustrialism' is obviously of a very different kind from the one of Bell or Kahn. It is a postindustrialism characterized above all by the basic criterion of sustainability, a concept which also has become central in the Another Development tradition (discussed in the previous chapter).

It is somewhat strange that the Counterpoint visions of postindustrialism are supposed to be 'utopian', whereas the Mainstream prophecies are taken to be 'realistic'. In fact 'utopian' has come to denote courses of change which substantially depart from the modernization trend. Since the universality and permanency of this trend now is fundamentally in doubt, it should be more appropriate to seriously consider alternatives, whether these are called utopian or not.

It is true that the revival of utopian social theory began in the 1960s, the decade of uninhibited consumerism, but, as Krishan Kumar notes: 'The problems identified and the values celebrated outlived the particular moment that may have driven them into public consciousness' (Kumar 1978: 260). Only by attending to the sorts of issues raised in the critique of industrialism is it possible to formulate strategies for sustainable development. One of the key concepts in this critique has been self-reliance, implying among other things a more domestic orientation of the industrialization process.

6.1.2 Self-reliance in Europe

In his efforts to apply development theory to European development problems, Seers was primarily referring to the European periphery: Portugal, Spain, Greece, Ireland. In order to grasp the development problems in these areas, theories originally developed in the Third World, for example dependency theory, should be of use, one reason being that economics as taught in Western Europe, on the whole has ignored hierarchical relationships:

> There happens to be a particular reason in the late 1970s for applying this approach in Europe. The discussion of the possible further enlargement of the European Economic community will come to a head in the near future. All three governments which have applied for membership (those of Greece, Portugal and Spain) lie on the periphery of Europe; so do possible future applicants such as Turkey and Cyprus. The incorporation of these countries into a predominantly core organization raises structural issues that are familiar in the development field. (Seers et al. 1979: xix)

The removal of regional inequalities and pockets of backwardness was one of the overall aims of the EEC. However, little has been achieved in terms of regional harmonization, partly because of the enormous dimensions of the problem, but also probably due to lack of theoretical understanding of polarized development in a European context. The early architects of the EEC were much too optimistic about the equalizing effects to growth, and they saw growth as more or less inevitable. This was in line with the academic theory of regional integration at the time. For this body of theory Western Europe subsequently became the main empirical case. Theorizing took different forms and went through several stages: functionalist, federalist and neofunctionalist (Harrison 1974).

Economic integration in Eastern Europe was seen, at least before 1956, as a completely coercive and exploitative process with little relevance for theory-building. If the Western European experience was rather unique, economic integration among centrally planned economies was unprecedented. The most remarkable feature of the emerging pattern was that of a dominant 'centre' exporting energy and raw materials to the 'periphery' in exchange for manufactured goods. Nevertheless there is great reluctance among the participants. In Etzioni's (1968) terms the Eastern European model is coercive, the Western European utilitarian, but what both seem to lack is the identitive source of integration.

Going back to the Western experience, it would be an exaggeration to say that the integration has met with great enthusiasm among the concerned populations in the six, then nine, now twelve countries. No one seems to want EEC, but the fear of being left out is obviously greater. The foreseen revolution of 1992 (the establishment of the inner market) is already making the remaining countries panic. The lack of authentic interest (outside Brussels) is significant. Perhaps it is the developmentalist vision of Europe as manifested in the Treaty of Rome that is the problem?

An 'alternative' vision, more realistic if the problem of spatial inequalities in Europe and on the world level is to be taken seriously, would be the one derived from the Counterpoint tradition: a more self-reliant Europe in symmetric cooperation with the Third World, and with new patterns of internal cooperation. In fact, self-reliance was a basic strategy in most European countries reacting to, but nevertheless pursuing, the industrialization path carved out by England (Senghaas 1985). In that particular historical context self-reliance implied the setting up of strategic industries with a strong military content and – at least during a period of transition – a protectionist policy *vis-à-vis* the external world. The similarities between the old and the new NICs are striking.

The alternative concept of self-reliance implies a permanent (if not static) situation rather than temporary withdrawal. Instead of specialization and maximum exchange, in accordance with what economists have termed 'comparative advantages', the emerging alternative tradition stresses a maximum of self-reliance for each natural economic unit or ecosystem, which does not necessarily coincide with the nation state. This view contradicts the Mainstream model, derived from European economic history. However, in no way would a self-reliant Europe imply an isolated and autarchic Europe. According to Stefan Musto: 'It implies a selective form of interdependence designed and shaped in accordance with the interests and potential of the territorial unit in question and its peoples' (Musto 1985: 6).

A key issue then is what we shall mean with a 'territorial unit' in this context. Self-reliance in Europe must be conceived of as a symmetric pattern of cooperation expressed on various territorial levels, and ultimately also in Europe's relations with the rest of the world. This can obviously not come about without some coordination within and between the nation states. Therefore it is necessary to formulate a general principle of territoriality, rather than conceiving development simply in terms of specific levels or regions. This principle is evident in the concept of ecodevelopment and in the distinction between territory and function.

As we emphasized in the previous chapter the revitalization of territorial life would not be possible without a transfer of power to local communities, while the state level takes the function of coordinator. It is also necessary to strengthen territorial units larger than nation states, for example the whole European continent.

In terms of functional v. territorial principles, the Mainstream Model is a rationalization of interests that have a central and dominating position in the functional system, mainly big business, politicians and bureaucracy, i.e. the power structure behind the European common market project. The legitimacy of those who control the commanding heights of the functional system has until recently been taken more or less for granted. Protests and disobedience from local leaders following the radically different logic of territorial systems were earlier considered as an act of political disloyalty. At present there is in many European countries a revival of 'localism', in that parliamentarians are increasingly experiencing a loyalty-conflict between their respective party line and a position consistent with the local interests of their constituencies. This is bound to have an impact on the party system.

The traditional political parties follow the logic of the functional system, i.e. maximizing economic growth mainly with reference to that system. Parochial interests normally had to give in to demands from the macrosystem, but today there are local parties with little significance outside their own territorial system, constituting a specific local community. Apart from these rather heterogeneous local groups, green parties are also emerging on the national level. They challenge the functional principle as a national development logic. This may be taken as a sign of crisis for the functional system, which has to grow continuously in order to compensate for dislocation, imbalances and strains created by the very process of growth. This 'cleaning-up' argument has today become the very rationale of growth. Since the conventional remedies merely worsen the situation, the current crisis also implies a strengthening of the political Counterpoint tendency, or what was earlier referred to as 'Third System politics'.

Without political change there is of course no chance whatsoever of the European model being transcended. Thus our argument rests on one basic assumption: a change of the European collective consciousness. Furthermore, a convergence of political processes arising in the 'first', 'second', and 'third' worlds is in fact necessary because of the present level of interdependence. Europe cannot change without changes in the rest of the world, and changes in Europe will at the same time have an important impact on the world.

6.1.3 Human needs and western man

If the reality of European development has been violent industrialization, the ideology of development has been brighter. From the ancient Greek philosophers to more recent development theorists of different persuasions the European conception of development has, as was analysed in Chapter 2, been identified with growth. In more recent times this growth process also became identified with 'progress' and, with the birth of development theory, 'modernization'. The Marxist tradition is no exception to this evolutionist bias, which also implied a paternalistic attitude towards non-European cultures on their way to 'modernity'.

Challenging this Mainstream paradigm there has also been a Counterpoint tradition arguing for the inherent superiority of small-scale, ecologically sustainable and decentralized patterns of development. This tradition is revived and strengthened in times of crisis for the dominant paradigm.

If self-reliance provides certain principles with regard to planning and economic-spatial organization, the Basic Needs Approach is more directly concerned with the goals and content of development. What relevance to an alternative European development does this approach have?

In the Basic Needs Approach two rather different lines of thought can be distinguished: one related to the Mainstream tradition in Western development thinking, the other an expression of the Counterpoint tradition. The first concept is marked by the politics of growth, redistribution and welfarism. It is materialist and universalist. A certain standard of living, below which no one is supposed to fall, is established. This idea is implied in the 'redistribution with growth' discussion, a social-democratic contribution to Western political thought. The second concept is expressed in the debate on quality of life, a debate which has often been described as confused and vague (Lederer 1980), probably a result of the fact that people tend to differ on what makes life worth living after 'das Fressen' is secured.

The first concept may not be irrelevant in the West where the level of production and consumption by no means can be taken for granted, particularly since some development strategies of recent popularity are threatening the material standard for marginalized groups. However, in the industrialized countries there is also a growing interest in the second line of thought, which tries to define human needs also in terms of social relations, freedom, security, participation, relations to nature, and even transcendental orientations.

If the structure of technologically advanced industrial capitalism is associated with 'politics of welfare', one legitimate question is of course why there is a revival of populist utopianism particularly in the West. Has Western Man reached the psychological limits of industrial civilization, or is there some premonition that the future viability of this structure is threatened by internal contradictions, or by external changes? On this we can only speculate, but it is a fact worth noticing that today there exists a large and growing literature on development alternatives in the industrialized countries. On the political level there are quite substantial social movements all over Europe for alternative lifestyles. ('Western Man' of course includes both sexes, but could also be interpreted as 'Western Male' in this context, since the Mainstream paradigm is marked by male values, whereas women are overrepresented in the new movements.)

It is interesting to note that these movements often combine ecological consciousness and the ideal of frugality with international solidarity and a commitment to the poorest of the world. These movements could represent a passing fashion (in my view unlikely), or they may be *groupes porteurs d'avenir*, i.e. people equipped with a particular foresight and the will to experiment with different solutions to perceived dangers.[81] Middle class intellectuals, representing the tertiary sector and typically female, tend to be overrepresented in these groups. They express a Counterpoint ideology, whereas both capital and labour normally opt for Mainstream solutions. However, as the high rate of unemployment in the various Western national economies stabilizes, the new lifestyle movements will increasingly recruit also working class participants looking for fresh solutions to the problem of survival. People are forced to take command of their own situation and cater for their basic needs, as the social safety net is being destroyed by economic depression and political neoliberalism.

Historically Western society was moulded by popular movements, which gradually modified the inequalities between classes and introduced democratic institutions. The traditional role of the labour movement was to defend labour against capital, but on the whole it accepted the 'mainstream' model of development, thus promoting further growth of the First and Second systems. The new social movements, on the other hand, articulate alternative values. It seems as if they converge on a fundamental critique of the predominant development paradigm.

Third System Politics can, as we noted above, be seen as a reaction to emerging problems related to the industrial system, the welfare state and the interstate system. The problems are many and of different kinds, and so are the political reactions. Thus the new social movements are typically issue-oriented; there are movements for alternative security, for alternative lifestyles, for alternative production, for environmental protection, for solidarity with the Third World (often specialized on different continents and even different countries), for women's liberation, and for regional and ethnic emancipation. In spite of the diversity and strong issue-orientation of all these movements it could be argued that they have something in common with the revolt against modernity in the Third World.

The new social movements are obviously inspired by various Counterpoint ideologies, such as anarchism, utopian socialism, and populism. However, the most important factors behind them are the dysfunctions of contemporary modernity: the overkill capacity of modern weapons systems, the civilization diseases, the ecological catastrophes and imperialism. Therefore there is a link between the antinuclear movements in the 1950s, the anti-imperialist movements in the 1960s, the environmentalist movements in the 1970s and the new peace movements in the 1980s.

There are also links between the new social movements and the rise of reciprocity, discussed in Chapter 5. This is because there are inherent values in reciprocal forms of exchanges, although these may not be sufficient reasons for bringing about the change. The basic factor is not the intellectual and emotional appeal of new ideas, but the future performance of the capitalist world economy. The recovery of the late 1980s does not absorb the unemployed in a significant way. It is possible that the process of marginalization will add strength to the 'counterpoint' or 'green' movement. It may in fact be the decisive factor determining whether these movements will remain middle-class-based, postmaterialist manifestations in the Western world, or if they will also become mass movements of unemployed, searching for 'useful unemployment'. In the next section we shall make a more concrete analysis of the emerging development alternatives in the European context and the competition between them.

6.2 Development options in western Europe

The nation state and the Mainstream Model of development originated in Europe but the whole world is now carrying the burden of this obsolete conception of development. Thus Europe has a moral obligation not only to preach alternatives to others, but also to provide an example by practising them. This would imply a retreat from the conventional European Model and a search for a new pattern of development, as also to find a new role for Europe in the world (Hettne 1988a).

Many Europeans today favour the idea of a European delinking: a Western Europe less dependent on the US, an Eastern Europe more independent of the Soviet Union, and the two Europes coming closer together. In view of the extreme dependence of most European countries on international trade, an increased level of self-reliance on a European level is exigent. However, self-reliance cannot be realized without alternative defence. For this reason peace issues and development issues cannot be properly understood in isolation from each other. There are however different ways in which the two can be linked (Hettne 1984). There are also different strategies whereby the two values of peace and development can be secured, since both are normative concepts with different meanings for different actors, trying to realize social projects.

By 'social project' I mean a specific course of action that contains a vision of the world, a strategy for realizing that vision, and a number of actors, working from a social base.

Unless we believe that history unfolds itself in accordance with some hidden law, we have to understand the future as shaped by struggle as well as alliances between different social projects. The concept of 'social project' transcends conventional social science analysis, whether it takes 'social class' or 'nation state' as point of departure, and tries to make justice to both territorially oriented local actors and transnational actors which are responding to world-system dynamics. The main rationale for this approach is in fact the necessity to understand contemporary political currents in a world context, since the conventional acting subjects in social science are usually associated with national arenas. Later, we shall deal with the new development ideologies emerging from this transnational imperative.

6.2.1 Alternative social projects

This section briefly presents seven more or less competing social projects on the contemporary European scene.[82] In the next I make a comparative analysis regarding the development strategies implied in the various projects, namely:

1. The Atlantic Project.
2. Peaceful Coexistence.

3. A European Superpower.
4. The Trilateral Project.
5. Global Interdependence.
6. Fortress Europe.
7. The Green Project.

These projects are seen from a European point of view, although their political arenas sometimes are smaller than Europe as a geographical area, sometimes bigger. They have, however, been chosen from the point of view of Europe, and we are interested in the specific role of Europe in the world, suggested by them.

The Atlantic Project and Peaceful Coexistence projects are in many respects similar, in spite of a fundamental geopolitical contradiction. Their points of departure are, in the former, United States national interests, and, in the latter, Soviet interests. Europe plays a subordinate role in both cases, but nevertheless forms part of the social projects through local collaborators, i.e. actors committed to either of the two.

Both imply continued superpower dominance and hegemonic struggle in world politics. Strategy, however, is only one side of the coin, the other being the economic alliance, established through the Marshall plan which was meant to protect the market economy and the open system of trade (Smith 1984). The crucial role of the Marshall plan for the subsequent development thinking has been emphasized earlier. The economic alliance was as unbalanced as the strategic one, the rationale of both in fact being to maintain the US hegemony in the postwar international order.

As for Peaceful Coexistence, it should be pointed out that this concept originally was used with reference to the coexistene of a large number of 'nations' within the Soviet state, but later on the concept has been used as a description of the Soviet position *vis-à-vis* the capitalist world. The concept therefore had to be open-ended, adapting to the constantly changing relationship. The concept postulates the existence of two camps with contradicting and competing socioeconomic systems. This systemic contradiction must not be allowed to manifest itself on the politico-military level. Instead it should be resolved through the ultimate victory of the inherently superior socialist socioeconomic system. In order to protect this system, as long as it remained vulnerable, the Warsaw Treaty security system was created. Within this system the Soviet Union exercises hegemonic power, similar to the US role in the NATO security system.

Recently the Warsaw Treaty was prolonged until 2005 with little internal debate among the countries concerned. It is evident that the pact is basically defensive, and that the belief in the ultimate victory of socialism now is growing thin. Therefore 'peaceful coexistence' has been reinterpreted as an intensified East–West cooperation in the economic and technological fields in order to make it possible for the socialist countries to catch up, or at least not be left further behind. This implies a relaxation of bloc politics, undermining some of the assumptions on which the project rests.

A considerably more ambitious project, as far as the role of Europe is concerned, is to restore Europe as Superpower. This prospect was much discussed in connection with the first enlargement of the EEC in 1973, but as was discussed above the speed of integration has been very uneven, oscillating between expansion and consolidation.

France has been the most enthusiastic nation state behind the Superpower project. Recent historical research on the origins of the Treaty of Rome (Milward 1988) disproves the popular myths about the early phase of European integration as a fulfilment of a long process of integration where national interests were subsumed under a prevailing spirit of Europeanness. Rather, the European Community agreements furthered the particularistic objectives of national economic reconstruction and the strengthening of the nation state. For France the EEC was instrumental in furthering her great power ambitions, whereas smaller countries like Belgium and the Netherlands saw the agreements as a guarantee for free trade conditions essential for their economic development.

France is consequently worried by the American SDI initiative, meant to consolidate the Atlantic alliance and the predominance of the US within that alliance. President Mitterand therefore initiated the so-called Eureka project, which is meant to be an all-European research project with partly the same R&D elements as the SDI, but with a clearer civilian orientation. All Western European countries (including the neutrals) have responded favourably to Eureka, which shows the extent to which the 'Pacific challenge' now worries the Europeans. However, the content of Eureka is still very much in the air, and it is not very likely that Europe will be a successful contender for world hegemony.

Compared to the three projects discussed so far, the Trilateral Project and Global Interdependence are of a more reformist nature, particularly in their approach to world order. They both stress the urgency of the development problem, defined in terms of decreasing demand and fragmentation of the world market. Since the world economic crisis is a threat to stability and peace, these social projects see disarmament and détente, in combination with economic liberalization on a global scale, as the political preconditions for peace and development. The concern for disarmament is perhaps more pronounced in Global Interdependence (as expressed for instance in the Brandt Commission reports), but Trilateral thinking as well, contains a commitment to détente and relaxed East–West relations.

The proposals from the Trilateral Commission can be seen as a response to the crisis from transnational capital. In principle they support some reformist elements of the Interdependence Project. The Trilateral philosophy thus also contains the idea of interdependence, the 'management of interdependence' being the key word. Here the socialist world is explicitly included. In the Trilateral conception, 'peaceful coexistence' would of course imply convergence of the two rival socioeconomic systems in the context of a liberal world economy, rather than the ultimate victory of socialism. The Trilateral approach to the Third World is on the other hand rather selective, showing an interest primarily in conservative regional powers and the dynamic NICs.

The Brandt Commission reports are silent about the role of the socialist world, while giving a strategic part (as a Keynesian demand raiser) to the South. In both projects, however, Europe is figuring more prominently than in those discussed previously, particularly so Global Interdependence, which to a large extent could be seen as the contribution of European Social Democracy to the struggle for a New International Economic Order. The Trilateral strategy implies the sharing of the hegemonic burden among the US, Japan, and Western Europe, whereas the world-view of the Interdependence project is more pluralistic. The role of disarmament as the main source of transferable resources is also a crucial part of the Brandt Commission philosophy. The Keynesian logic behind this global strategy for peace and development is obvious.

The somewhat militant concept of Fortress Europe was coined by Wolfgang Hager (1985) in the context of identifying a proper unit of economic management in Western Europe. According to Hager the EEC is at present not such a unit, if the spontaneous market forces are taken into consideration. Furthermore, due to the recent enlargement the internal heterogeneity of the EEC will further increase. This reinforces the need for internal and external industrial management, 'which might drive a wedge between the EEC and countries with identical socioeconomic systems remaining in the so-called rest EFTA'.

The idea behind Fortress Europe is to take advantage of the 'domestic' market and combined productive capacity of a unified Europe (the EEC plus what is left of EFTA, and, in a longer perspective, possibly even the Eastern European countries). This would create a large economy which, like the US and Soviet economies, could be basically self-sufficient. Protectionism on the European level seems to be the only way to save a unique European achievement – the welfare system – since the alternatives (protectionism on the level of the nation state, or monetarism à la Thatcher) are paving the way for economic anarchy, destabilization, social conflicts, and – ultimately – political tensions.

In a recent book the late Dudley Seers (1983) stated: 'The last quarter of this century is making mercantilists of us all – really the only question is what to protect: the nation or the continent?' Although sceptical about mutual interests in what is here called the Global Interdependence project, Seers saw such mutual interests within a greater Europe as more realistic. He argued for a strong integration with efficient redistribution policies based on 'extended nationalism' in order to eliminate European underdevelopment. In a large perspective he also foresaw a closer cooperation between Eastern and Western Europe, complemented by nonaggression pacts that would make both NATO and WTO obsolete. Such a development should also make it possible for the neutral countries to join Europe, which in itself would become neutral in the process of Europeanization.

European self-reliance would be the logical counterpart of South–South cooperation. A world system with a number of strong regional groupings would be safer and more stable than the present world system. This neomercantilist vision does not exclude interregional links. Andre Gunder Frank's pan-European strategy, for instance, includes European ties with the Middle East and Africa. However, these interregional relations would be different from the vertical relations in the present hierarchical world system (Frank 1983a).

It is obvious that such a restructuring of Europe necessitates a revival of state interventionism and economic planning in order to avoid an imperialistic structure within the European system. This is consequently a project of the European left. According to the 'Out-of-Crisis' project of European socialists, there are three keys to European recovery: reflation, restructuring and redistribution (Holland 1983). These policies have to be carried out on a European level, since unilateral expansionism will defeat its own purpose, as the French example showed.

The Green Project is now attracting supporters in most European countries; more, it seems, in the north than in Latin Europe. Green parties are growing stronger in local bodies, but face certain difficulties in nation-level politics. 'Die Grünen' in West Germany are obviously the most active and articulate group and are often seen as the 'mother party' by others. There is no Green International since the Green movement (which one should see as something broader than Green parties) in itself is transnational due to the issues involved. Which then, are the main issues?

The first issue is environmental protection, which according to the Green view is incompatible with the modern industrial system, whether socialist or capitalist. Rudolf Bahro, who has experience of both, borrows E. P. Thompson's concept 'exterminism' as a general characterization of Western civilization (Bahro 1984). A worldwide industrialization would destroy life on earth. Before that happens, however, the competition for dwindling resources would increase the level of political tension in the world system to a point when a nuclear disaster would destroy the world. Thus the second issue mobilizing the Greens is peace. Ecology and peace are linked, but their realization needs a transformation of society.

The problems that activate the Greens are therefore different from the usual campaign issues. Ecology and peace constitute fundamental starting points to which all other issues must adapt. They are not just a few points in a lengthy party programme, nor are they merely issues in an election campaign. This is why the Green Project deserves some attention in this context. The share of votes that the Green parties get is from this point of view less important.

Green politics is a variety of Third System politics. As the Third System grows (in the form of informal economic systems, networks rather than state-controlled macro-organizations, stronger and more autonomous local communities, etc.) the role of the state (First System) will automatically diminish, as will the large corporations (Second System) operating in functional rather than in territorial systems. From the point of view of the giant corporations the revitalization of territorial life is a market loss, just as it represents a tax loss from the point of view of the state apparatus.

6.2.2 Comparative analysis of development strategies

Before discussing the development strategies of the seven projects we must say something about their overall development goals and world-views. The first three projects (The Atlantic Project, Peaceful Coexistence, and A European Superpower) share a rather conventional world-view. According to this view, the world essentially constitutes an anarchic interstate system where each state is a sovereign unit, controlling its territory, counting upon the loyalty of its people or 'citizens', always prepared to defend its territory and people against all other nation states.

Every nation state is also a functionally organized national economy, competing with other national economies on the world market. In accordance with their comparative advantages they become specialized, which means that goods are produced and distributed in an increasingly effective way. This ultimately benefits all the participating states, even if the short run effects happen to look a bit gloomy. The specialization creates functional links between states, supporting tendencies towards peace, and countering the inbuilt tendency towards conflict.

Obviously the conventional model does not quite operate in this way. The fundamental contradiction in the model is between the logic of the world economy and the logic of the international political system.

- The Atlantic project seeks to maintain the US postwar hegemony as a necessary precondition for 'the free world', threatened by Soviet expansionism.
- The Coexistence project implies a dualistic world order, with the Soviet Union exercising hegemonic power over the socialist world.
- The European Superpower project, in contrast, transcends the nation state system, but essentially conceives Community Europe as a kind of nation state – a third superpower with ambitions to take a hegemonic role.
- The Trilateral and Interdependence projects are less state-centric, since they conceive the world as an interdependent system, a system which can be reformed precisely because of this interdependence. The vision, based on a consensual model of society, presupposes a significantly increased commitment on the part of nation states to joint political management of the system of global interdependence, as a functional equivalent to the role of hegemonic power in the conventional world order. The Trilateral Project, however, renders a key role to organizations such as the IMF and the World Bank, while the Interdependence project relies on political cooperation within the UN system. This is an obvious contradiction between the two projects, which can be seen as right or left varieties of the vision of Interdependence respectively.
- Fortress Europe articulates a different vision: a regionalized world, consisting of a number of independent, self-reliant blocs. Each bloc would constitute a natural ecological and cultural region, large enough to be basically self-sufficient.
- The Green Project contains an even more far-reaching vision of a decentralized world system, with strong self-reliant local communities, a sustainable society where basic human needs are fulfilled, and where the relationship between society and nature is nonexploitative. This vision firmly negates not only the nation state, but also the liberal world order.

Development strategies are sometimes discussed as if there were a free choice between them; as if they were different means to achieve an end. However, development strategies do not simply express the 'Zweckrationalität' of men in their pursuance of the all-embracing goal of development – they are indistinguishable from the very goals pursued. There are different development paradigms, and the development strategy follows from and is marked by the principal approaches to development, derived from different paradigms.

In this context it is again helpful to recall our distinction between functional and territorial principles of development. The former corresponds to the mainstream pattern of development: economic growth resulting from specialization and division of labour between regions. What ultimately counts is the aggregate result on the level of the national economy, or, according to some visions, even the world economy. The problems of regional underdevelopment are either seen as temporary imbalances or necessary sacrifices in order to achieve overall growth.

The territorial principle, in contrast, gives priority to regions and local communities. Their development must be consistent with the ecological and cultural characteristics of each region. It is development from bottom up.

The predominant paradigm of development is functionalist, and therefore most of our social projects, albeit with different focuses, adhere to the functional principle (see Fig. 6.1). One important difference concerns the roles to be played by the state and the market; another one to what degree the national interest should be subsumed under the logic of the world market, or the imperatives of interdependence.

- Contemporary Atlanticists are at the same time nationalists and strong believers in the magic of the market. Their nationalism is anticommunist since communism is (or at least it has traditionally been) antimarket.
- The Coexistence project is not only nationalist but also statist. However, socialist self-reliance has led to stagnation and therefore the socialist countries have tried a certain degree of world–market integration, which in turn again has underlined the state-market contradiction.
- Superpower Europe is market oriented and 'nationalist' on the level of Europe, but Europe is primarily seen as a base for the penetration of other regions by European capital. Thus the development strategy is functional rather than territorial.
- The Trilateralists and Interdependentists are ready to give up a substantial amount of national sovereignty for the benefit of functional development on the world level. They differ on the

Fig. 6.1 The relation of the seven social projects to state–market and the functional–territorial principle

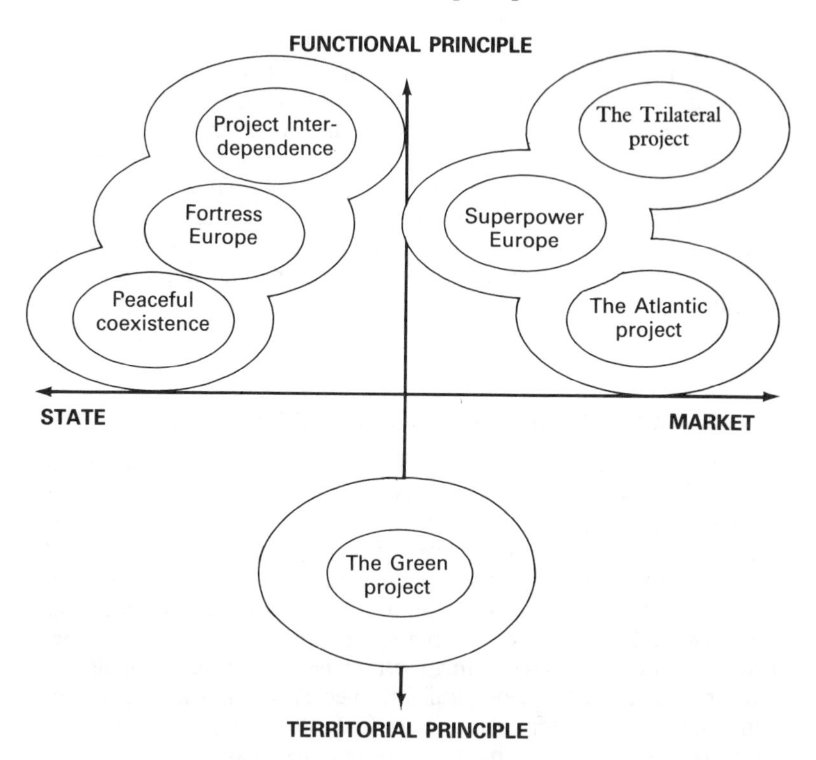

need for political intervention in the economic process. The interdependentists accept intervention in support of the European welfare state and propose the following strategy for global welfare:

First, the level of armament should be lowered, but with a sufficient degree of deterrence maintained, since the anarchy assumptions still are held to be valid. Second, the resources released by the politically agreed on reduced spending on armament should be transferred to the developing countries, so that the purchasing power of these countries may be increased. Third, the growing.demand resulting from this massive transfer of resources will supposedly invigorate growth and employment in the crisis-ridden industrial countries.

We may perhaps call this worldview a social democratic world order, because of the obvious similarities to a national economy organized along Keynesian lines.

- Fortress Europe and the Green Project reject integration on the world level. According to Bahro 'there is no salvation without dismantling this complexity', whereas proponents of Fortress Europe see the problem in the combination of a complex division of labour with the anarchic market conditions. The Green Project emphasizes territorial development with a local focus, more or less in line with the Another Development tradition. It is strongly antistatist. Fortress Europe is more interventionist and more positive to functional development, but wants the process of integration to make a full stop at the level of Europe. Within Europe the political power centre would have the main responsibility for controlling the market forces and removing regional inequalities.

6.2.3 Contemporary development ideologies

In Chapter 2 we discussed development ideologies in European economic history. The liberal model, state capitalism, the Soviet model and Keynesianism were described as Mainstream manifestations. Chapter 5 was devoted to several historical Counterpoint manifestations, among them populism. We shall now see what emerging contemporary (and future) development ideologies can be derived from the previous discussion on social projects and their development strategies. Four such ideologies will be discussed: neoliberalism, neo-Keynesianism, neomercantilism and neopopulism. The first three are more or less Mainstream, the fourth a contemporary expression of the Counterpoint. In comparison with the traditional list of ideologies the absence of socialism on the new ideological map is conspicuous. It is true that world-system theorists foresee a transition to a socialist world system, but no ideology of world socialism is elaborated. However, there are socialist elements in neomercantilism. This discussion is limited to Western Europe, while the next section takes a look at the East European scene where the ideology of socialism as well is undergoing remarkable changes.

By adding 'neo' to the old ideology's name my point is to emphasize continuities as well as discontinuities. The old and the new forms are similar enough to go by the same name but there is a distinct new element in the neologism. I would say there are two important new elements in all the four development ideologies which we are going to discuss here:

- The 'neoideologies' are reactions to, and reflections of, the current crisis.

- Whereas the old ideologies were articulated within a national space and took the national interest more or less for granted, the neologisms relate to a global space. They represent various stages in a process of transcending the nation state logic and becoming world development ideologies.

Neoliberalism

Long ago liberalism was born as an expression of freedom from the absolute state and a defence of the market against 'unnatural' mercantilist regulation. It was also a defence of the individual against the state, and from that point of view almost indistinguishable from certain anarchist trends in the eighteenth century.

In the course of its development liberalism recognized both a social responsibility for all citizens of the state (J. S. Mill) and a role for the state in the management of capitalist development (J. M. Keynes). This development, later reinforced by organized labour and the formation of social democratic parties, laid the foundation of the welfare state in Europe.

Neoliberalism is basically a reaction to the contemporary welfare state, particularly the power structure constituted by labour unions and state bureaucracies, accused of having destroyed the market system. This is how the neoliberals explain the crisis, and the way out is consequently to dismantle the welfare state and return to the market.

The best summary of neoliberalism is Milton Friedman's: 'To the free man, the country is the collection of individuals which compose it. . . The scope of government must be limited . . . to preserve law and order, to enforce private contracts, to foster competitive markets . . .' (Friedman 1962: 2). It is not very surprising that the modern industrial countries find it hard to accept this message and to abandon a well-established economic doctrine, but most of them are nevertheless moving towards the market pole of the mainstream axis. Most determined in this regard have been the Thatcher and Reagan administrations, whereas a country such as Sweden, due to the identity of the 'Swedish Model' with the Social-Democratic interpretation of Keynesianism, is slow to reconsider policies that have been so successful until now.

Neoliberal ideas have been around for quite some time and the main spokesmen besides Friedman have been Ludwig von Mises and Friedrich A. Hayek. When Keynesianism was declared dead, neoliberalism got its long awaited opportunity. Furthermore, it became part of the new right movement which among other things resulted in Thatcherism and Reaganism. Thus neoliberalism is one strand in the New Right (Levitas 1986) but could and should be analytically separated, since neoliberals do not have to be domestically authoritarian or externally Cold War warriors. However, this is the context in which the ideology usually appears:

> The real innovation of Thatcherism is the way it has linked traditional Conservative concern with the basis of authority in social institutions and the importance of internal order and external security, with a new emphasis upon reestablishing free markets and extending market criteria into new fields.
>
> (Hall and Jacques 1983: 121)

Neoliberalism is obviously a reaction to the crisis (particularly the crises of the welfare state), so our first criterion is fulfilled. What about the second? Is neoliberalism a 'world development ideology'? Obviously, the liberal economic doctrines always claimed to be universally applicable and the free-trade doctrine is of course, if anything, a world development doctrine. On the other hand the current neoliberal manifestation as part of the New Right, with its parochial and nationalist overtones, seems to be contradictory.

In the last decade a new development orthodoxy has also preached 'liberalization' in the Third World. This is the 'counter-revolution' in development economics (Toye 1987a), which will have important implications for development thinking. One example is the monetarist school, which significantly has re-established itself in Chile, the former stronghold of not only the dependencia school but also the CEPAL tradition (see Chapter 3). Monetarism primarily refers to a rather technical discussion regarding causes of inflation, but in the context of development strategy it has implied an extreme laissez-faire approach, as in this declaration of the Chilean Junta: 'Our concern is to transform Chile not into a nation of proletarians but into a nation of proprietors'. This is akin to what in Thatcher's England is called 'popular capitalism'.

At the Cancun top meeting on global development problems in October 1981, the neoliberal philosophy was translated into 'global Reaganomics', while the Keynesian strategy contained in the Brandt report (which originally was meant to provide a framework for the discussion) was tacitly buried.[83] Instead the developing countries were advised to liberalize their economies, encourage their entrepreneurs and find out their comparative advantages. This recommendation implies a return to the ideal model of capitalist development from which actual development in the industrial countries themselves has deviated substantially. Nevertheless, it is also the current policy of the IMF and the World Bank. All this made monetarism the dominant paradigm throughout the 1980s.

Neo-Keynesianism

If Keynesianism implies the management of economic fluctuations and the objective of establishing the balance between supply and demand through the instrument of the state, neo-Keynesianism applies this logic to the world-economy as a whole. Keynesianism fell into disrepute early in the current crisis, a common explanation being that Keynesian policies did not work because the national economies had become more or less dissolved due to the increasing independence in the world economy. Thus both our criteria for adding the prefix 'neo' are fulfilled.

Neo-Keynesianism claims to have remedies for the crisis but they can only be applied in a global context. The Keynesian logic behind this strategy for peace and development is most explicitly dealt with by Angelos Angelopoulos:

> . . . the great problem today for the industrial countries is to generate a 'creative demand' . . . It is therefore necessary to find sufficient purchasing power for absorbing additional goods produced. If such purchasing power cannot be secured within the national borders of a country, it must be sought elsewhere. The developing countries, which have a great need for equipment and services for accelerating their economic and social progress, have sufficient levels of unsatisfied aggregate demand but lack of the requisite purchasing power. Here is the key of the problem. And it is here that the reapplication of the Keynesian theory on a world scale can bring a solution to the present economic crisis by generating sufficient levels of global creative demand . . . By interpreting the Keynesian theory in this global way, Keynes can be considered as the precursor of the 'new world economic order'.
>
> (Angelopoulos 1983: 139–41)

We have earlier discussed the neo-Keynesian logic of the Brandt Commission reports, and the same approach can be found in the proposals coming from the Socialist International. Global Keynesianism derives its attractiveness from simplistic assumptions that the world is fully integrated through the market principle, and that trade is unambiguously beneficial. To the extent that these assumptions do not hold true, the effects of a global Marshall plan may be extremely varied.

Neomercantilism

Mercantilism is the ideological expression of the nation-state logic operating on the economic arena, violating the liberal principle that free trade in the long run is for the benefit of all. Neomercantilism retains the suspicion that the old mercantilists harboured against free trade, but like neo-Keynesianism it transcends the nation-state logic in arguing for a segmented world system, consisting of self-sufficient blocs large enough to provide 'domestic' markets and to make use of economies of scale and specialization in production on the one hand, but without falling prey to the anarchy of world capitalism on the other.

Such a regionalized world system would probably be more stable than a liberal world order which has revealed an inherent tendency towards collapse. Each region in a regionalized world system would be large enough to have a reasonable degree of economic efficiency, in accordance with the principles of comparative costs, economies of scale and other conventional economic arguments. On the other hand social perversions generated by excessive specialization and an overly elaborated division of labour could be avoided.

There is a difference between this new form of protectionism and the traditional mercantilist concern with state-building and national power. What we could call neomercantilism is a transnational phenomenon. Its spokesmen do not believe in the viability of closed national economies in the present stage of the development of the world economy. On the other hand they do not believe in the viability of an unregulated world economy either. Nor do they – in contrast with the Trilateralists and the Interdependentists – put much faith in the possibility of managing such a world economy. Rather the neomercantilists believe in the regionalization of the world into more or less self-sufficient blocs.

This is the 'benign' view of mercantilism summarized by Buzan (1984: 608) as follows:

> The benign view sees a mercantilist system of large, inward-looking blocs, where protectionism is predominantly motivated by considerations of domestic welfare and internal political stability. Such a system potentially avoids many of the organizational problems of trying to run a global or quasiglobal liberal economy in the absence of political institutions on a similar scale. The malevolent view sees a rerun of the mercantilist dynamic of the past, in which protectionism is motivated primarily by considerations of state power.

Thus the benign view is essentially the same as what I here call neomercantilism. It must be emphasized that this 'neomercantilist' vision faces serious problems of acceptance due to the strong historical association of mercantilist thinking with extremist nationalism, and because of its periodical revivals (in the terms of protectionism) in connection with economic crises. This 'guilt by association' argument can however be rejected, since no real efforts have in fact been made to construct a world order on the basis of such mercantilist ideas. The established view that such ideas tend to re-emerge in the context of world depressions and collapsing world orders could not be held against them, for the simple reason that constructive propositions must be distinguished from desperate responses in a situation of crisis.[84]

Neopopulism

Populism was essentially a defence of the territorial community against the functional system created by modern economic growth, both in its original capitalist form and its derivative socialist form. Neopopulism is an attempt to recreate community as an offensive against the industrial system, which is seen as a global machine threatening mankind with ecological destruction and nuclear holocaust. It negates mainstream development and in this negation lies the essence of the utopian vision, historically expressed by the counterpoint tradition in Western politics, as well as in the resistance against the imposition of the nonindigenous nation state on the peoples in the non-European world.

In the European context neopopulism is manifest in the fundamentalist currents within the Green movement, most explicit in the writings and political activities of Rudolf Bahro who has been a rebel in two socioeconomic systems – and for the same reason: the extreminist trend in modern industrialism, whether market- or state-organized.[85]

One manifestation of neopopulism is the Green ideology discussed earlier as one of the counterpoint manifestations and in this chapter in terms of the Green Project. We can therefore be brief here. Neopopulism resembles classical populism in several respects: the urge for community, the stress on primary production, the distaste of industrial civilization. However, there are significant new elements of relevance in this context: an environmental consciousness, encompassing the global ecological system, and a strong commitment to a just world order. The corresponding historical movements were more parochial.

Let me illustrate by again quoting Denis de Rougemont, who opened this chapter:

> Prospects for a worldwide solidarity of autonomous peoples, the sole alternative to global economic disaster and nuclear war, are linked to prospects for a federation of ethnic, ecological, and civic regions in Western Europe, as a living example of a postindustrial model capable of freeing the peoples of the Third World from their fascination with mechanistic stakhanovism.
>
> (de Rougemont 1980)

In this quotation all the crucial ingredients of neopopulism are contained: the fundamental critique of Western industrialism as a system, the emphasis on local transformation (including ethnodevelopment), and the global implications of a restoration of territorial communities.

6.3 The rise of market ideology in the east

So far we have discussed 'European' development issues mainly from a Western perspective. It is of course unacceptable to identify Europe with the western part of Europe. This section therefore focuses on the reality and ideology of development in its eastern part. The reality is that a development model, assumed to be the correct socialist development path, was imposed on a group of countries with extremely different preconditions for development. The ideology, whether expressed on the political arena or in academic form, should be understood against that background. This implies among other things that the context for socialist reforms differs radically between the Soviet Union and the other Eastern European countries.[86]

6.3.1 Soviet development thinking

Soviet development thinking is generally neglected in textbooks about development. In a broad overview like the present it is not possible to make up for this neglect. On the other hand, without any attempt to place the Soviet and other Eastern European theoretical contributions in a proper historical context, it would not be an overview in the first place. We shall first concentrate on three major debates in the Soviet Union, and later on widen the perspective to the rest of Socialist Europe. Third World socialism is on the whole left out of this discussion.

The crucial historical context is the centre–periphery structure within Europe, the old cleavage between East and West, and the modernization imperative for the eastern latecomers, most of them characterized by weak state structures and large agrarian populations. In this perspective one must understand also the Russian revolution, the proper starting point for surveying the Soviet debate, although many of the issues had in fact been discussed long before.

The classic issue, and the first great debate, was whether Russia in order to develop had to go through the ordeal of capitalism, or whether socialism could be built on the foundation of traditional agrarian collectivism which formed part of the Russian tradition and national identity. The Russian populists or narodniks, whose views were outlined in the previous chapter, provided the framework for the development debate from the latter half of the nineteenth century and well into the 1920s, when their ideas, considerably weaker at that time, mingled with the second great debate, the one about industrialization. The first debate has been called the 'controversy over capitalism' by a major authority in the field (Walicki 1969).

The question was if, and under what conditions, a country in a precapitalist stage could make a direct transition to socialism. It is easy to appreciate the extent of the controversy, since the narodnik position implied a strong element of voluntarism, challenging the Marxist view of orderly transition from one mode of production to another. What makes the debate so exciting is that Marx and Engels were actual participants. Their position was somewhat delicate since they from a purely theoretical point of view were sceptical with regard to the possibility of bypassing capitalism. On the other hand they did not want to discourage the emerging revolutionary movement in Russia. They avoided the dilemma by linking the possibility of direct transition to socialism with the condition of a proletarian revolution in Western Europe. It is of interest to note the similarity between this argument, closer to political tactics than to development theory, and the more recent Soviet theory of noncapitalist development. As we remember, this theory states that underdeveloped countries can circumvent, or at least shorten, the capitalist stage, on the condition that they receive support from the socialist bloc.

Going back to the original debate, it was closed by Engels, after the death of Marx, in favour of the determinist position, which in Russia had been taken by Plekhanov and later was developed by Lenin (Lenin 1967).

The narodnik position is, as we earlier emphasized (see also Kitching 1982), the most articulate of counterpoint manifestations. The Russian populists denied the possibility of capitalist development in Russia – not the development of capitalism in Russia, but the development of capitalism as a nationwide mode of production. They looked upon the efforts of the Russian government to introduce capitalism as a 'parody of capitalism'. Russian capitalism had no external markets, and at the same time it could not produce for internal markets, since the development of capitalism, by bringing to ruin peasants and artisans, reduced the purchasing power of the population (Walicki op. cit: 117). Russian capitalism was doomed to develop towards what much later was discussed as 'peripheral capitalism'.

The Russian populists thus for the first time raised the peculiar problem of backwardness coexisting with advanced economic systems and the special problems and possibilities connected with development in this context, i.e. the problem of contemporary underdevelopment as analysed by the dependentistas. The similarity between the Narodnik positions and contemporary Third World ideologies has created a renewed interest in populism among Soviet scholars (Khoros 1984).

The first of our three debates was a Mainstream-Counterpoint debate. The second debate, usually referred to as 'the Soviet industrialization debate' (Erlich 1960), which raged during the 1920s, was a Mainstream controversy. At the same time it was a political struggle between spokesmen of NEP, most importantly N. Bukharin, and those arguing in favour of rapid industrialization. The discussion concerned the prospects for the socialist project in a backward country under conditions of isolation, and particularly focused on potential sources for the accumulation needed for the industrialization process. One such source was the peasantry existing outside the system of socialist relations of production. E. Preobrazhensky, the most important leftwing theorist, in his 'The New Economics' from 1926, used the controversial concept 'primitive socialist accumulation' (Preobrazhensky 1965), whereas his still remaining Narodnik opponents spoke of 'internal colonialism', referring to the exploitation of the peasants necessarily implied in industrialization.

Students of the Soviet industrialization debate agree that it is extremely rich and can be seen as a rehearsal for the development debate after the Second World War, concerned with the Third World.[87] Therefore this account has to be schematic, focusing mainly on the positions of Bukharin and Preobrazhensky.

Defending the socialist instrumentality of market exchange in the Soviet Union has been politically controversial as it could be criticized either as a restoration of capitalism or as a rehabilitation of purged comrades (or both). A recent case of rehabilitation is that of Bukharin. Gorbachev's many references to Lenin's later works might in fact be understood as references to Bukharin and the latter's apology for the New Economic Policy (NEP) of the 1920s. Bukharin's development strategy was of course a strategy to build socialism through the methods of the NEP, not a defence of the class interest of the kulaks (as he was accused of). It was in fact also a strategy of industrialization, but at a 'snail's pace', with the workers–peasants alliance intact, and without excessive centralization of state power.

The problem with this strategy, as was pointed out by Preobrazhensky, was that it did not provide for accumulation. His own solution, 'primitive socialist accumulation', would on the other hand have endangered the fragile alliance with the peasants. The two positions expressed a development dilemma, later to be 'solved' by Stalin. The inherent contradictions in this type of solution were bound to preserve the legacy of Bukharin as a 'political undercurrent' in later Soviet economic debates (Lewin 1975).

What Bukharin stood for was in brief: distrust in excessive state power, balanced growth, the view of peasants as allies in the project of building socialism, and market socialism. These principles are not less relevant now, although both the context and the content of them have undergone important changes. Together they make up a rather different socialist project than the 'really existing socialism' of today. This theoretical tradition in the field of economic planning does not of course consist of the ideas of one single person, but was elaborated by a group of nonparty economists whose work is of course well known in the Soviet Union (Judy 1971; Jasny 1972). This body of work is now bound to become further revived.

When the 1920s came to a close the industrialization strategy originally developed by E. Preobrazhensky carried the day, imposed as it was by Stalin and the modernization imperative, concretized by the growing threat from Nazi Germany. According to Preobrazhensky the mixed socialist economy of the 1920s was dominated by two 'laws': 'the law of value' and 'the law of primitive socialist accumulation'. Since they were incompatible and the Soviet economy by definition was socialist, the first, corresponding to the market mechanism, was bound to disappear. This theoretical position conceived socialism as a totally planned economy, the main purpose of which was accumulation under state control. It had to be 'primitive' accumulation because of the low level of the development of productive forces when the Russian revolution occurred.

Behind the positions of Bukharin and Preobrazhensky, articulated on the political level, there were hosts of brilliant economists participating in what they conceived of as a theoretical discourse. Stalin meant the end of the debate – and the end of analytical economics (Judy op cit: 218), or at least its suppression until the 1950s.

Before turning to this awakening, it should be emphasized that the debate of the 1920s is probably the richest that has taken place in the field of development economics, preceding the birth of the subject by decades. As we noted in Chapter 2 there were several Eastern Europeans among the pioneers in development studies. The Soviet debate, to which should be added the spectacular results of the Soviet industrialization model, exercised a strong influence on the early development of development economics, and to some extent explains the emphasis on industrialization and the bias in favour of interventionism.[88]

The third great debate on socialist development thus concerned the question 'What beyond Stalin? It started immediately after Stalin's death, and was of course not less 'political' than the previous debates. Thus there were two issues involved: how to mend the imbalances built into the dirigiste framework, and how to remove power holders who continued to have vested interests in this framework.

The specific economic problems were three: the lagging agriculture, the underdeveloped consumer industries, and the lack of flexibility inherent in the Stalinist planning model. The trial-and-error process associated with attempts to solve these problems is to a large extent the history of the rise and fall of Kruschchev, which we cannot go into here. Suffice it to say that the most interesting debate took place between 1962 and 1964, and was dominated by E. G. Liberman who wanted to 'free central planning from detailed supervision over enterprises, from wasteful attempts to control production not by economic but by administrative methods' (as quoted in Nove 1964). Rather than administrative decentralization (as in Krushchev's aborted sovnarkhoz-reform) the Liberman economic philosophy emphasized the firm as the crucial unit of decision-making and implementation. This implied different economic signals such as 'profit' and the demand–supply mechanism. Thus the old issue of 'market socialism' reemerged from history.

The Soviet economic reforms in 1965 were rather bleak compared to the foregoing debate and should be understood as a closure of the debate rather than as a courageous step. What followed was rather an era of conservatism later to be dramatically interrupted by Gorbachev. Even before that, Soviet economists had begun to question the relevance of the 'classic' model for developing countries. Instead NEP-type policies were seen as more appropriate in the context of underdevelopment. Such policies would include: considering agriculture the basis of capital formation; promoting light industry to provide employment and income for the peasantry; undertaking measures to ease agrarian overpopulation; building up labour-intensive industries; creating a domestic industrial complex; engaging in extensive economic and technical cooperation with the advanced states (Kridl Valkenier, 1980: 495).

In 1985 a new economic strategy was proclaimed, based on the three pillars of 'uskorenie' (acceleration), 'perestroika' (restructuring) and 'glasnost' (openness). The new economic strategy is related to a new deal in international relations, stressing concepts and perspectives (interdependence and common security, for instance) strikingly similar to the views of the Brandt Commission and the Palme Commission, of which Georgi Arbatov was a member (Aganbegyan 1988; Arbatov 1983). The years which have passed since this announcement have been dramatic and the repercussions all over the socialist world are fundamental – much of it probably yet to be seen.

Under Gorbachev the relevance of NEP not only for developing countries but for the Soviet Union itself was increasingly discussed. The ghost of Bukharin became more and more evident until the not unexpected rehabilitation in early 1988. In Eastern Europe, particularly in Hungary and Poland, the anti-Stalinist reformers had to some extent been inspired by Bukharin's writings. Thus, as Lewin pointed out with reference to the first and second waves of reforms, opposing camps were formed along ideological lines similar to those of the 1920s (Lewin 1975). Even today the crucial issue is how to combine planning and market – and the dilemma how a decentralized economy can be combined with a party monopoly of political power. That this cleavage has become relevant in the other Socialist countries as well is due to the diffusion of the Soviet model, to which we now turn.

6.3.2 The imposition of the Soviet model

From 1948 to 1953 the Soviet model or near replicas of it were adopted by all communist regimes in what was to become 'Eastern' Europe. (For a summary outline of the content of the model see Chapter 2.) Its most significant characteristic was the inward orientation; the emphasis on what later became known as self-reliance. Stalin's concept of 'socialism in one country' implied the construction of a centrally planned (coercive) national economy which had to be sufficiently comprehensive and differentiated to provide for almost all the needs of society. This implied a pervasive role for the state.

Another feature which should be kept in mind is that the socialist development strategy was meant to be more than a development strategy per se, namely a strategy for socialist transformation (Brus 1987). I have referred to this as the 'project of building socialism'. Thus it cannot be assessed only in terms of economic criteria but also with regard to achievements towards that particular goal.

In the Soviet Union this goal was established as a result of the revolution (although there were competing interpretations of socialism in the postrevolutionary debate). In Eastern Europe the goal had never been articulated on a broad political basis. Furthermore, the project of building socialism was subordinated under Soviet security concerns. The indigenous roots of socialism were weak, and socialism as such by many conceived of as imported or imposed, in spite of the relaxations after 1956.

For this reason the prospects of the socialist development strategy in Eastern Europe to succeed, either as a strategy of development or as a strategy of socialist transformation, were never great. In terms of economic development the record is mixed, and in terms of the socialist project as such (on which the opinions are bound to be subjective) there are evident reasons for disappointment. The most spectacular case of imminent failure right now is Romania.. Popular protests against socialism should be a contradiction in terms, even if they are directed against the 'actually existing' socialism and not the socialism that could be.[89]

As far as economic development is concerned, it is true that an impressive industrial base for production was established in countries where no such base previously existed, and that growth figures on the whole compared well with Western Europe, at least under the period of 'classic' industrialization. The improvements on the side of consumption have been less impressive, 'shortage' being the most common experience for the Soviet and East European consumers (Kornai 1980). Since the rationale of postponed consumption in the socialist development model is to facilitate accumulation, the slowdown in economic growth without notable improvements in living standards over the last two decades is fatal, as is now realized all over Socialist Europe (Bozyk 1988).

The Socialist countries find themselves in some kind of structural trap. The reasons have been aptly summarized in the concept of 'conservative modernization' (Brus op. cit: 153), containing problems such as:

- Technology: The imitative technical progress does not produce a spillover effect in spurring home grown technology (apart from military-related sectors).
- Structure: The traditional leading sectors (such as steel and heavy engineering) have lost their role without being replaced by the new industries.
- Foreign trade: The trade with the West bears the mark of underdevelopment with a predominance of primary goods.

To this must be added social ambitions (inherent in the project of building socialism) which are incompatible with modern norms of efficiency in production (which tend to be set in the West). The most crucial one is the principle of full employment, now abandoned in the West but until recently an integral part of socialism in the East, 'where the workers don't expect to get paid, but on the other hand are not expected to work either', as the joke goes in many of these countries. The kernel of truth is that work is badly paid but reasonably safe and most workers do not see any need to overexert themselves. An enormous amount of work is in fact quasiwork (hidden unemployment) and huge resources could therefore in principle be saved by shutting down enterprises, if someone dared apply the axe.

Finally there are fundamental political problems in the 'ancien regime', which are best dealt with in terms of legitimation (Fehér, Heller and Márkus 1983: 137 ff.). On this point generalizations are unhelpful since the political traditions in the various socialist countries differ considerably. What unites them is that they are monoparty systems based on the principle of 'democratic centralism', i.e. the ruling party is ruled by a ruling elite. What differs is the degree to which this arrangement is accepted, i.e. the degree of legitimacy. If the reforms are objectively necessary, as was indicated above, they can be implemented from above to the extent that the principle of 'democratic centralism' is widely accepted. If not, the reforms will only be implemented if a new basis of political legitimation can be established.

This is the background to the current politics of reform in the East, including the Soviet Union under Gorbachev. However, the necessity of reforms does not mean that they automatically will be implemented.

6.3.3 The dilemmas of reform

The imposition of the Soviet model ruled out the possibility of a more 'organic' growth of a socialist society in Eastern Europe – to the extent that such a possibility ever existed – and led to what today generally is referred to as 'actually existing socialism'.[90] Since the socialist model was imposed rather than being indigenous to Eastern Europe, the reform movement is as old as the model itself. Being fundamentally political, i.e. concerned with the distribution of power, it has been a fluctuating process: the three waves of the early 1950s, mid-1960s and 1980s (Brus 1979).

The events in Prague 1968 were a major setback for the reform movement in the East and perhaps destroyed the prospects for the socialist project beyond repair. There are reasons, however, to be more optimistic about the present (and third) reform wave in the East, simply because this time the Soviet Union itself is taking the role of the avant-garde. Paradoxically, conservatives who once established themselves through Soviet support are now challenged by 'the Gorbachev effect'.

What, then, is the meaning of reform in the Eastern European context, and what reasons are there to believe that reforms provide solutions to the socialist development problem? For a summary of the need for reform I can do no better than to quote Dieter Senghaas:

> The lack of efficiency in the economy, which results from a hypertrophic planning bureaucracy, from no longer expedient priorities and probably even now from inadequate division of labour among socialist societies, prevents adequate satisfaction of accumulated consumption wants and thus undermines the economic legitimation of socialist development.
>
> (Senghaas 1985: 197)

The dominant recipe in the debate on socialist reforms has been modifications of the coercive model through the introduction of some sort of market signals, influencing resource allocation between alternative productive activities.[91] Ever since the NEP this has been seen by orthodox Marxists as a retrograde step in the transition towards communism, and proponents of reform could therefore easily be accused of being harbingers of capitalism, which is also precisely what has happened. As emphasized above, there is a political, as well as an economic side to this issue. Economic reforms imply changes in the power structure and are therefore resisted by internal as well as external actors.

Thus there are so far limited experiences of the effects of economic reforms – more exactly we have three cases, Yugoslavia, Hungary and Poland, on which conclusions may be drawn. There can, in view of present performance, be no overall positive conclusion about the effects of these reforms. The lessons to be drawn must rather rest on a detailed analysis of the actual function of the reforms in their historical contexts. What was the nature of the problems to which they were to provide a solution, and under what constraints did they operate?

The Polish crisis, dealt with in the first chapter, must be understood against the background of a series of earlier unsuccessful reform attempts, the most important 1957–60 and 1971–73 (Wilczynski 1986). The reasons for the failures are inconsistencies, ambivalence in implementation and conservative opposition – in combination. Our concern here is with the more recent debate on market socialism, or, differently put, the transition to developed (mature) socialism (Bablewski 1987). Maturation implies the creation of a socialist welfare state with a balance between different sectors as well as between production and consumption priorities. In Poland the characteristics of this transition period were widely discussed in the 1970s and onwards. The concept of an 'open socialist economy' came to be seen as a break with the inward looking legacy of socialist development thinking (Plowiec 1978). The concept implied a planned economy participating in the international division of labour and the use of external resources in the accumulation process.

In the 1980s Polish theorizing went strongly against the directive type of planning. It stood out as obvious that previous attempts at reforms had not significantly touched the 'system of early 1950s' (Hübner 1988). Decentralization of the management system, restructuring of industry and export-orientation were emphasized. Theoretical inspiration also widened from Marxist approaches to 'catholic' and 'populist' (Bablewski 1987), which of course implies quite different theoretical perspectives on the very function of reform. Even socialist theory is less than clear on this point. From a normative point of view, central planning – which became identical with central management – was established as the authentic socialist model. Thus 'the state became the manager of a multiplant enterprise called "national economy"' (Wilczynski op. cit.). A contradiction between plan and market was assumed and it was the goal of socialist transition to overcome this contradiction.

The Yugoslav and Hungarian experiments constitute distinct models with specific institutional solutions: workers' self-management and market socialism in agriculture. These models contain their own contradictions and in evaluating them it is important, but difficult, to try to distinguish problems inherent in the socialist transition from problems which must be explained by the historical heritage.

Take for instance the serious economic and cultural disparities between the Yugoslav provinces and the chronic unemployment in that country (Bidelux 1985: 182). Even with a well-managed economy Yugoslavia would face recurring political crises stemming from the ethnic diversity of the state formation. Thus the question of reform is inseparable from the federal arrangement and the development of interethnic relations, which in turn spill over into the sensitive area of national security. Of importance in this context is also the uneven development towards political pluralism in the provinces, a process where Slovenia is taking the lead.[92]

The two most serious economic problems in the Yugoslavian economy are inflation and unemployment. Both are linked to the logic in which the self-management system operates – if not caused, they are at least aggravated by it. As Nove (1983: 138) points out there is no material interest in taking on extra labour if that means that the net revenue per worker is reduced. Similarly private enterprises (employing not more than 10 persons) can only expand through labour-saving methods if they want to remain private. As for inflation there is a contradiction between the logic of the overall economy and the logic of the individual firm, where a rise in productivity is linked to a rise in pay, even if it is cancelled out by other production costs imposed by external conditions. To resolve these contradictions one has either to revert to central planning (a difficult proposition in the Yugoslav federal context) or to take further steps in the direction of a market economy (which is the more likely course). Either way, the anarchy of 'Balkanization' is a distinct and threatening possibility.

The Hungarian reforms were introduced in 1968 and although they removed some of the problems associated with the central planning model, the New Economic Mechanism was soon facing contradictions of its own, e.g. between social security and economic efficiency. Efficiency on the microlevel may however not coincide with efficiency on the macrolevel, something which is difficult to see from the point of view of the single firm. Thus the implementation of the reforms was inconsistent due to political interventions, which created confusion. Nevertheless the reforms must be seen as successful until the international crisis of the 1970s imposed constraints which further exposed the internal contradictions.

As our examples should have made clear, the need for reforms is economic but the means will have to be political as well, implying fundamental changes in power relations. This is one dilemma of reform in Eastern Europe.[93] There must, on the other hand, also be some self-imposed constraints in the general praise of the market mechanism if the socialist project shall be able to maintain some credibility and legitimacy. The present reform policies of some Eastern states raise the problem how the distinctiveness of socialism as an economic system shall be conceived. As reported in *Herald Tribune* (7 April 1987) Leszek Balcerowics, reform theorist at the Polish Central School of Planning and Statistics, has said: 'The dream of an economic system better than capitalism is dead.' Janos Hoos of the National Planning Board in Hungary is quoted as having said: 'We have to clarify the matter of principle. If we continue to proceed pragmatically we will fail to resolve the question of what is socialism and how it is different from Western capitalism.' In the Soviet Union itself the economic debate was polarized, one school drawing arguments from the NEP of the 1920s, and another defending central planning as an 'indisputable achievement' of socialism (*Washington Post* 9 June 1987).

On the level of ordinary people the main issue of course is whether reforms mean any improvement in their living standard or not. As always, any change would have differential effects on various groups. One group, already vulnerable, is the hidden unemployed in the huge bureaucracies and the many inefficient enterprises referred to above. Their number could well be something like twenty per cent of the working force, thus exceeding the worst cases of unemployment in the West.[94] *The Economist* (26 Dec. 1987) is probably right in saying that 'one of the many threats to the Gorbachev revolution is the possibility that people will in the end prefer the charming stability of the old inefficiency to the pain of perestroika'.

The reform movement in Eastern Europe has a cyclical character, which probably is related to the unwanted political consequences of far-going reforms. The economic system is also a power structure and therefore no economic changes are possible without political changes modifying this power structure. As a matter of fact, much economic reformism is political opposition in disguise. What better argument could there be against a fossilized centralism than highly idealized descriptions of how a free market operates, such as the ones found in Western economic textbooks?

Thus, the reforms have their excesses and their backlashes. The first reform wave started after the death of Stalin and ended abruptly in Hungary 1956. The 'second wave' started in the mid 1960s and continued until the early 1970s, although the backlash of this wave came already in 1968 in Czechoslovakia, paradoxically at a time when Hungary entered a phase of the most radical experimentation so far. The 'third wave' is the current one started by Gorbachev, and if there is to be a backlash this time as well, it will probably come in the Soviet Union itself. However, the backlash may be provoked by excesses in Eastern Europe.

The three waves are not uniform in all socialist countries. On the contrary it is quite natural that individual countries react differently depending on whether the reforms (including the political ones) were blocked, reversed or consolidated in the previous phase. Hungary was slow in the second phase due to events during the first, but is today in the forefront because of the consolidation of the reforms which took place in the second phase. On the other hand Czechoslovakia is now lagging behind because of the nonreformist political reorganization which took place in the post-1968 period. The 'third wave' seems particularly problematic since the economic and political reforms are now seen as components of the same package.

On the contemporary political scene the ideology of socialism seems to be a spent force. In the First World socialist movements face increasing difficulties in conquering state power, in the Second socialism is often defined as a part of the problem rather than a solution, and in the Third the socialist periphery seems to be generally worse off than the capitalist. It is unavoidable that antisocialist forces are in a jubilant mood. However, a historical perspective is essential in making judgements on societal experiments. Socialism and capitalism are not systems which can be applied to real situations like pills, and then compared with respect to their curative power. Rather they are, in spite of simplistic labelling, so many societal situations indistinguishable from the historical processes preceding them.

I have argued that the historical function of socialism has been to organize transitions to capitalism in backward areas. This, however, does not mean that a future function of socialism may not still remain. As long as the system of production and distribution is irrational from the points of view of human needs and ecological sustainability, the socialist tradition will always constitute a source of criticism and utopias.

Chapter 7

Reorientations in development theory

Development is still a contested concept, and this is perhaps as it should be. Instead of providing a synthesis, this concluding chapter therefore focuses on discontinuities and divergencies, as if we were dealing with a 'discipline' in disintegration. Whether this is actually the state of affairs is a matter on which opinions necessarily diverge.

I maintain that the career of development studies is intellectually exciting, and that it has brought back important and forgotten issues in social science theorizing.[95] It is, however, hard to see this intellectual process as an accumulation of wisdom. More appropriately, one could speak of an accumulation of 'social science sects' (Gareau 1987). This process can be conceived of as one single theoretical project, even if its participants do not appear to pursue a common cause, and many babies are carelessly thrown out with the bath water.

The present study is an attempt at outlining the birth, evolution and transformations of this nebulous branch of social science: development theory or, less solemnly, development studies. Its main concern is theoretical changes, but the pertinent question remains: changes of what? Paradigms in the social sciences tend to accumulate rather than replace each other, leading to a proliferation of social science sects. Thus, to those who continue adhering to the earlier development doctrines, simply because they find nothing fundamentally wrong with them, no paradigm changes have occurred. Instead, development studies in many recent publications from right and left is dismissed as a digression (Bauer 1981; Lal 1983), or as having reached an impasse (Booth 1985). I do not agree with the stronger denunciations and tend towards a defensive position. This discussion of development studies is consequently subjective – but not as subjective as attempts at synthesis would be. Such attempts are currently being made, however, and should to my mind be welcomed as breaks with exaggerated polarizations and as necessary, albeit provisional, stages in the development of a more integrated discipline.[96] Whether this is a possible task in the long run is a different matter. In this context I shall merely summarize what I see as major reorientations in the field, some of them increasing the diversity, but some also creating common ground for consensus.

I will discuss four such reorientations, grouped in two sets. The first is cyclical, and results from shifting positions along two dimensions with a long tradition in the social sciences:

- the positive-normative dimension;
- the formal-substantive dimension.

The second set constitutes secular trends, or what I conceive of as more or less irreversible processes:

- the change from a Eurocentric to a contextual and pluralistic understanding of development;
- the change from endogenism, over exogenism, to globalism.

The following sections summarize the shifts in theory with regard to these dimensions. A concluding section discusses the future of development as a policy concern, in view of the various reorientations.

7.1 One field or many?

In this survey different theoretical approaches have been brought together on the perhaps somewhat arbitrary assumption that they jointly form an emerging tradition. One valid objection to be faced therefore is that the concept of 'development theory' might have been stretched inordinately, so as to accommodate too many unrelated phenomena. Is there, for example, any connection whatsoever between issues such as the transition from feudalism to capitalism, the lineage mode of production, capital accumulation, terms of trade, sustainability and quality of life, just to mention a few concepts which have been used by development theorists of various persuasions. Should all these be regarded as forming part of a development theory and – if the answer is yes – to what degree could we claim that we are dealing with a reasonably coherent intellectual tradition?

A look at recent introduction texts from various disciplines reveals that the theoretical discussions usually are uniformly organized around the modernization v. dependency cleavage, occasionally with a chapter on 'alternative' or 'another' development added. Secondly, it is repeatedly asserted that the geography, sociology or politics of development do not make sense as isolated, mono-disciplinary approaches.[97] This trend has strengthened my conviction that development studies must be seen as a single, inter-disciplinary field.

Such a trend may, however, also be interpreted as a broader

change in the social sciences towards fully incorporating the development problematique, thereby creating a more general convergence towards a unified social science and completely transcending development studies. For instance Foster-Carter (1985: 1) has observed: '. . . really it is the sociology of development which should include all the rest of sociology, rather than the other way around.'

There are thus at least three different interpretations of the career of development studies:

- It was a nonstarter, confusing the issue of development, which essentially amounts to the same thing in all countries.
- It is a new, still emerging, discipline dealing with the special circumstances of underdevelopment in the Third World.
- It is a revival of the developmental orientation in classical social science, but based on a broader global experience.

It is not possible to tell which interpretation is the correct one. Probably all three will continue to have their champions, unless one of them assumes the monopoly of definition. Personally I disagree with the first; I have for a long time based my work on the second; but I am beginning to believe that the third interpretation is the most reasonable. This book, dealing with 'three worlds' rather than the Third World, is a step in that direction.

In what follows I will argue that the research territory so far has been defined by an ongoing debate between positions which have moved along the positive–normative and formal–substantive dimensions, which I have borrowed from more or less well-known Methodenstreiten in the social sciences in order to put the methodological controversies within development studies in perspective.

7.1.1 The positive–normative dimension

In ideal-typical terms positive studies deal with the world as it is, normative as it should be. Most research actually contains a bit of both. What is of importance is the basic approach. The distinction between positive and normative analysis has a long tradition in Western intellectual history. It is well established in economics but is perhaps particularly controversial in International Relations theory. One confrontation between the two modes of thought within that field came already with the Renaissance, when Machiavelli suggested that, at least for princes, it might be better to study man as he really is, rather than as he ought to be, as was done in the earlier, mainly theological, debate on man and society. The 'realist' way of thinking emerged victorious and later the concept of 'interest' (in contradistinction to the 'passions') stood out as a new paradigm (Hirschman 1978). The idea that people tend to be guided by their (material) interests culminated in the concept of economic man, which provided a base for a positive science of economics.[98] The ideal of a positive, value-free social science survived many attacks and reigned supreme after the Second World War.

Development theory, which was a post-war phenomenon, is by its nature implicity normative, but became explicitly normative only in the 1970s. This is what usually is referred to as normativism.

From a dialectic point of view, there is no reason for expressing judgements about either approach as inherently good or bad. The wave of normativism in development studies was in part caused by previous excessive positivism, guilty of triviality and irrelevance. For instance, the view of development as inherent in all societies and going through foreseeable stages was both trivial and irrelevant, since the specific preconditions for development in various regions never were explored in an unbiased way. Thus, there has been a recurring tension between a tendency to study development as it actually takes place in the light of preconceived theories, and a tendency to theorize about what development should imply according to declared values.

On the whole, however, there has been a strong bias in the social sciences against research that explicitly utilizes preferences and values in the definition of problems to be investigated. During the 1970s the dominant trend was normative and voluntaristic, but with the more recent counter-revolution 'realism' seems to be taking the initiative again. This more recent change is supported by the immediate urgency of many contemporary development issues: debt crisis, pollution, food shortage, etc. The whole development agenda has in fact been changed.

Underlying this changing debate is the very definition of development, an issue which I purposely have avoided by calling it an 'open-ended concept'. Obviously the opening up of a concept towards greater inclusiveness gives more room for normative thinking about what development should be.[99] For a theoretical survey this conceptual inclusiveness is essential, unless we want to exclude contributions which claim to belong to the field, thus favouring a particular conception of development. For other purposes a more precise definition of development is probably useful, but hopefully not to the extent of returning to the old identification of development with growth.[100] Thus, on this dimension the positions move back and forth, ultimately correcting each other, even if each swing may appear exaggerated.

Most nontrivial social science theories are, however, based on certain values. Without degenerating to wishful thinking, development theory should not fear to be explicitly normative, critically evaluating ends and means, rather than searching for a hidden conformity to law in what actually takes place. The way I understand some present trends, they are pregnant with unpleasant consequences and mankind would be very unfortunate if these could not be averted. There should therefore be no incompatibility between the study of what is and the theorizing about what ought to be as long as a distinction is maintained. Furthermore, the 'ought to be' questions do not have to be treated with complete subjectivism. There is no fundamental contradiction involved.

Obviously the ideal society a particular development theorist dreams about is of limited interest to others, particularly if we do not share his values. It is, however, quite feasible to formulate normative principles with a reasonable degree of acceptability. One principle could for instance be adherence to certain ethical values, such as respect for 'inner' and 'outer' limits of development. Secondly, the way of solving a problem should be reasonably durable, i.e. one should avoid a medicine that helps the patient today, only to kill him tomorrow. Thirdly, it should be consistent with the way other societies solve their problems so that problems are not exported instead of being solved, i.e. the 'beggar my neighbour' syndrome.

There is a risk that the 'new realism' in development studies will breed cynicism. The principles of Another Development, discussed in Chapter 5, have not lost their relevance – on the contrary. The problem, however, is to mobilize sufficient political strength behind them. This may be done through conscientization, or – the slow and painful way – through the lessons of history.

7.1.2 The formal–substantive dimension

The second methodological dimension contrasts the formal approach, where development is defined in terms of a limited number of universal goals and quantifiable indicators which can be combined in a predictive model, with the substantive approach, where development involves historical change of a more comprehensive, qualitative and less predictable nature.[101] In order to grasp this perhaps somewhat elusive distinction, which also must be understood as an ideal-typical one, it is helpful to recall the not so recent debate among economic anthropologists regarding the general applicability of formal economics, particularly in its neo-classical shape. The substantivist crusade against economic man started with Malinowski's complaint about 'this fanciful, dummy creature', whose 'shadow haunts even the minds of competent anthropologists' (Malinowski 1961: 60). This line of criticism culminated with Karl Polanyi, whose life work may be summarized as a methodological critique of the false universality of economics, the creation of what he and his followers called a 'substantive' conceptual framework, and the application of this framework to different historical and cultural contexts (Polanyi 1957, Dalton 1968). Among influential defenders of the formalist approach, deriving economic change from a limited number of basic and universally valid principles, such as scarcity, choice and maximization, one may mention Firth (1939), Belshaw (1965) and Burling (1962).

In development theory more or less the same issue was raised in Dudley Seers's classical piece The limitations of the special case (1963), which argued against the universalist position taken by, for example, Robbins (1932) and Bauer (1957). The former rejects the view that the laws of Economics are limited to certain conditions of time and space, and the latter is quoted as saying:

> I am now convinced of the very wide applicability to underdeveloped countries of the basic methods of approach of economics and the more elementary conclusions stemming from these. (op. cit: 15)

Seers took a position similar to that of Polanyi in the formalist–substantivist debate in economic anthropology, which implied that the formalist (or universalist) approach really reflected a 'special case' (the logic of the market system). In different terms (used in the later debate on indigenization) it was a matter of academic imperialism. As is the case with indigenization, substantivism – the remedy to formalism – is not easily defined. Since it has to account for qualitatively different situations the degree of generalization must be reduced to the level of middle range theorizing, and modelling has to be satisfied with models for different types of economy. Economics becomes the study of economies. Theorizing must also take different political, social and cultural contexts seriously – not only as 'noneconomic factors'.[102]

The broadening of development theory from simplistic growth models towards more or less comprehensive and holistic theories of historical social change can be analysed along the dimension of formalism v. substantivism, also reflected in the repeatedly made distinction between growth and development (Arndt 1981). Already the step from economic growth to economic development is full of implications, and to see development in its political, social, cultural and ecological dimensions is an enormous enlargement of the research agenda – besides, as was pointed out above, inviting a great deal of normative reflection.

It should be emphasized that substantivism is different from particularism. It tries to develop theoretical frameworks that make full justice to particularities. Thus the substantivist challenge has been an important stimulus for the interdisciplinary trend in development research, whereas the formalist position of defence and, more recently, attack has been to stress the need for rigorous theory. In the early phase in the development of development theory the economic dimension was supplemented by other social science aspects, and subsequently the analysis was not only broadened but also deepened toward a historical–structural perspective. Furthermore, it was realized that development implied a process involving both society and environment. The breakthrough of this insight took place in the early 1970s. During the latter half of the 1980s most development agencies were again stressing this in their general programmes. Thus the ghost of scarcity did not only return – it came to stay. Formal theories and models therefore also tried to incorporate environmental concerns, but the issue whether there is a basic incompatibility between formal economic theorizing and ecological thinking is still unresolved.

In order to have a framework relating the diverse and often contradicting theoretical approaches to each other let us combine the two dimensions discussed above. In this way four major theoretical orientations are established:

- positive–formal;
- normative–formal;
- positive–substantive;
- normative–substantive.

The purpose of such a framework is not to suggest a taxonomy, but rather to locate the cyclical movements within the theoretical space delimited by the two dimensions. It is, however, difficult to find strictly comparable units in this type of analysis, since one single theoretical tradition in itself may contain different orientations in terms of the chosen dimensions. Various Marxist contributions, for example, must be put in different boxes. Some (for example the Althusserian structuralists) are clearly positive, others (for example the Frankfurt school) are far on the normative side, while Marx's own ambition was to transcend this dichotomy altogether (by understanding the world in order to change it). Nevertheless, Marx's polemic against all kinds of utopianism (called 'adventurism' in political terms) was so vehement, that his legacy on the whole has tended to favour positive rather than normative theorizing, as is particularly clear in the 'neoclassic' Marxist revival in development theory.

The classic Marxist model of accumulation, and more recent studies on the 'logic of capital' in particular, exemplify the positive–formal approach, whereas the mode of production analysis, as developed by Marxist students of the Third World, is considerably more substantive, however without abandoning the positivist (and typically Marxist) ambition to demonstrate that what happens in societies is somehow also necessary (Booth 1985: 773).

On the other hand the neo-Marxist school as well as radical dependency theory represented a movement in the normative–substantive direction. The normativism was evident not only from the political voluntarism, which characterized most dependentistas, but also from their emphasis on self-reliant development as an ideal. The substantivism was typically shown in the broad historical–structural approach of dependency.[103]

Drawing on the dependency perspective, world-system analysis retains the substantive orientation towards holism, but the normativism is in retreat and the positive orientation is again strengthened. Thus, the emergence of an interdependent world appears in this approach as a kind of natural history, providing little scope for national experiments with self-reliance, at least as long as the world system as such has taken no definite step toward socialism. Individual socialist countries operate under the same basic laws as the capitalist countries. Within the world-system approach, however, there are both formal (stressing a single global dynamics of the world system) and substantive (stressing the uniqueness of each world economy cycle and the openness of the future) tendencies. Consequently its location along the formal–substantive dimension depends on which position one takes.

Neoclassical growth theories, derived from the Harrod–Domar model, are obviously formal and – to the extent they prescribe rather than describe a development path – normative, for instance when they are used as planning instruments in developing countries. This normativism was usually implicit. The discussion on redistribution with growth and the basic needs approach in the 1970s was a step in a substantive direction, while the normativism was made more explicit. The counter-revolution again reaffirmed the universal relevance and validity of the market principle and, as a corollary, the imperative of 'getting prices right'. Structuralist economics, on the other hand, kept a distance to neoclassical analysis by insisting on historical specificity, institutional conditions and structural relations. It has stimulated several other theoretical traditions, among them dependency.

Compared to conventional growth theory modernization theory was broader in scope and its normativism less explicit. This may certainly be disputed, since modernization obviously implied certain development ideals derived from Western experience. Nevertheless, the typical modernization theorist held that the modernization impulse was inherent in all societies. The more recent 'soft' modernization position, conceiving modernization as a phase in rather than an attribute of history, is more positive and has a certain affinity with the world-system approach in some respects.

Fig. 7.1 A tentative summary of orientations in development theory

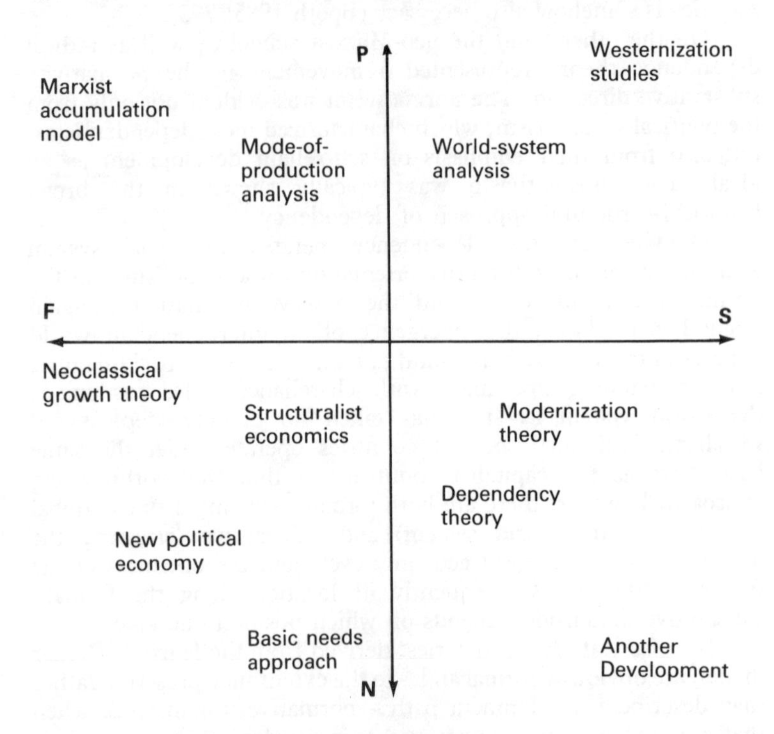

It is, however, much more on the substantive than on the formal side. We could call this 'Westernization studies' to distinguish it from the classic evolutionary modernization theory, or the 'natural history' approach. One example is the analysis of the expansion of Europe and its role in creating an international society as a distinct historical phase but without the theoretical underpinnings of the world-system approach (Bull and Watson 1984).

Another Development, which in the interpretation given here includes concepts such as self-reliance, basic human needs, sustainability and ethnodevelopment, represents a significant trend in the normative–substantive direction. According to the theorists of the alternative school, development should be an endogenous process, in this respect bearing a certain resemblance to classical modernization theory. On the other hand, different societies are, according to the alternative school, to follow their own development paths, inherent in their history, ecology and culture, an assumption which more or less rules out formal theorizing on a high level of abstraction. This body of thought stands somewhat isolated, which, as I see it, is due to the 'underground' Counterpoint tradition in Western development thinking. It emerges periodically as a protest movement against 'modernity' (or Mainstream).

The positions referred to here are tentatively located in Fig. 7.1. As I have understood the intellectual history of development theory, much of the debate must be related to the tensions between these dimensions to be comprehensible. This is what gives development theory a certain unity or 'residual common sense' (Preston 1985: 159) in spite of all its diversity, and this is also why I insist on treating development theory as one indivisible, albeit not very integrated, intellectual field. Most shifts do not imply a reversal to the exact previous positions but contain an element of learning, i.e. we are dealing with an upward spiral rather than a cycle.

7.2 Transcending Eurocentrism and endogenism

Let us now turn to the secular trends, which, in contrast with those dealt with above, I consider as more or less irreversible. Ethnocentrism is a dangerous kind of innocence which, however, tends to disappear with widening horizons. Eurocentrism should be no exception to the rule. Thus Europe will cease to be the standard by which development in other parts of the world is measured. Furthermore, development can no longer be seen as so many national developments realizing an inherent developmental logic. Instead, for each country the developments in other countries must be seen to provide the developmental context, representing possibilities as well as constraints. The transcending of Eurocentrism (via indigenization) is here referred to as universalization, the transcending of endogenism (via exogenism) as globalization.

7.2.1 The process of universalization

A major theme of this study is the transformation of development theory from its original Western parochialism to a whole range of interconnecting theories, reflecting worldwide experiences of different kinds of development problems. No General Theory of development has appeared and perhaps never will. What this study has shown is:

- Western social sciences are on the whole less universal than normally claimed by the gatekeepers of conventional wisdom. This has become particularly clear as mainstream development thinking spreads into other cultural spheres.

- The process of indigenization of social science in non-Western societies implies an intellectual emancipation as well as a fundamental reassessment of the Western paradigm. This, in turn, has given rise to a more self-critical attitude among many Western scholars, further enforced by emerging development problems in the industrial countries themselves.
- The twin processes of reconsideration of established paradigms within the West, on the one hand, and indigenization of development thinking in Third World countries, on the other, constitute the basic points of departure for the emergence of a truly universal conception of development.

This possible intellectual trajectory can be restated in terms of thesis, antithesis and synthesis. As thesis we may then see the global diffusion of Western development theory as a modern form of culture colonialism. The attempts at indigenization constitute an antithesis, which then, according to the well-known dialectical formula, would precede the synthesis: a universal concept of development. If the first process is referred to as westernization, and the second as indigenization, we may term the third universalization, a process transcending Western mainstream social science and articulating more diverse experiences (see Fig. 7.2).

The three processes thus constitute one single dialectical movement, in which extensions in application lead to a questioning of the validity of the theory and, through more or less radical modifications, stimulate more universal formulations. As soon as the 'limitations of the special case' were recognized, efforts towards indigenization were made, providing a base for further univer-

Fig. 7.2 The universalization of development theory

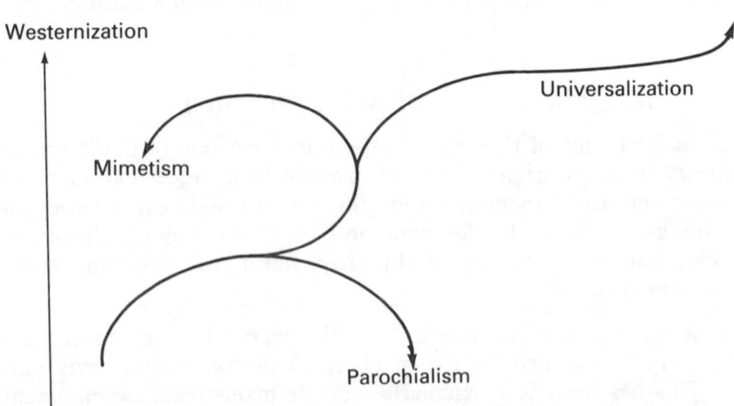

salization. Consequently, theoretical innovations, which originally were generated in the search for explanations of underdevelopment, are now increasingly being applied to the 'special case' itself.[104]

The notion of universalization is of course rather speculative, and critics will say that I have fallen into the Hegelian teleological trap.[105] Here I am, however, not making a forecast but rather outlining a project for development theorists. The ultimate goal of universalism may never be achieved. It could well be argued that along with the process of universalization, there are at least as many examples not only of continued intellectual penetration, leading to mimetism, but also different fundamentalistic rejections of westernization, leading to various new forms of parochialism. The process of universalization would at least make development theory more comprehensive and sensitive to contextual factors. Somewhat paradoxically, the universalization process therefore reflects the specificity of development. Indigenization in fact stands out as a precondition for universalization. Substantivism, which we discussed earlier, can be seen as a methodological response to the challenge of indigenization and thus a step towards universalization.

The ultimate test of the thesis pursued here would be whether development theory, enriched by its confrontations with diverse development problems in various parts of the globe, can throw new light on development in Europe, where development theory once was born as a paternalistic view of the 'backward' areas. The contemporary debate in Europe focusing on state versus market, glasnost and perestroika, democratization (in East and South), dangers and possibilities of interdependence, development and environment, adjustment and restructuring, the search for new 'leading sectors', the choice of technologies, and many more questions, all representing fundamental development issues rather than routine matters in economic management. Even if such convergencies of problems do not provide a base for a general development theory, the number of issues suitable for comparative cross-national research within a universalized development theoretical framework constitute an interesting challenge.

7.2.2 The process of globalization

The second secular trend in development theory is a corollary to the first, since transcending Eurocentrism also implies transcending endogenism, i.e. the idea that all countries in their development basically repeat the 'original transition', or the European industrialization process. According to the classic modernization paradigm development was seen in an evolutionary perspective, and the state of underdevelopment defined in terms of observable differences between rich and poor nations. Development implied the closing of these gaps by means of an imitative process, in which the less developed countries gradually assumed the qualities of the industrialized nations.

Modernization policies were thus not primarily seen as a development strategy, but as the working out of universal historical forces, resembling the transition from feudalism to capitalism in Western economic history. Among the modernizers there were fundamentalists who conceived of development as a basically repetitive and endogenous process, realizing the potential inherent in all societies in more or less embryonic form. The original Marxist view of development was largely shaped by this Eurocentric idea of Progress. New, higher relations of production (e.g. socialism) could according to this view not appear 'before the material conditions of their existence have matured in the womb of the old society'. The notion of 'underdevelopment' did not exist in the classical Marxist system, where the more developed country showed to the less developed the image of its own future. The development problem was simply to catch up.

The dependency position meant a complete switch from endogenism to exogenism, meaning a change in the search for explanations in domestic conditions to the international system of division of labour. The criticism of dependency theory consequently to a large extent concerned its overemphasis on external factors, but it should be kept in mind that the dependentistas rightly attacked the modernization theorists' obsession with internal factors and formulated their position in conscious opposition to modernization theory. Thus there have been two kinds of bias in development theory: endogenism and exogenism. Both approaches are, if carried to their extremes, equally misleading. The obvious remedy is to transcend this dichotomy and find the synthesis. There are in fact no countries that are completely autonomous and self-reliant, and no countries that develop (or underdevelop) merely as a reflection of what goes on beyond their national borders. Furthermore, the repercussions of the economically dominant countries on the developing countries are both positive and negative. This is true also for the previously secluded socialist countries as they enter the world market. Thus the 'three worlds' are forming part of a larger transnational system.[106]

We have noted that this basic fact often is described in terms of 'interdependence', a concept that lends itself to different interpretations. To some it is a refined form of dependency theory, expressing a more complicated structure of the world economy. To others the idea of interdependence suggests a common predicament for the peoples of the world, an interpretation which largely fulfils an ideological purpose. For this reason the concept of interdependence must be considered an ambiguous innovation. Nevertheless, several of the postdependency approaches, such as Marxist analyses of the internationalization of capital, neo-structuralism and world-system analysis were all marked by the necessity of grasping the dynamics of world development. This globalization trend therefore seems irreversible. However, we seem to be trapped somewhere between an obsolete 'nation state' approach and a premature 'world' approach, and the spokesmen of the respective approaches are prone to exaggerate their case.

The interest in global theories can be regarded as an effort to go beyond dependency in order to create a framework in which both centre and periphery, as well as the relations between them, are considered. However, whereas indigenization was characterized by cultural diversity and pluralism, globalization of development theory has so far not only been extremely structuralistic but also marked by a neglect of the cultural dimensions of the emerging global communications system. World-system analysis, for example, will have to account for the cultural implications of shifts in the geographical location of centres, as well as the mobilization potential of cultures still intact and only partially penetrated by Western values. The future of the world system cannot simply be derived from historic trends or some revealed inner logic which eludes empirical research.

By internalizing the external factors world-system analysis in a way was a return to endogenism of a more grand scale, the endogenism of the world-system rather than the endogenism of the nation state. Thus, the repeated attempts at transcending endogenism have not quite succeeded in escaping from the teleological trap. Early development theory took national development as a more or less automatic process for granted and was concerned with the barriers to this process, rather than the mechanisms behind it. To some extent this bias has been overcome, but still it needs to be stressed that development is a differential result of human action, and that any development process consequently can be reoriented through alternative human actions. The imperative to take the global context into consideration does not change this. To get rid of the air of metaphysics and determinism that still surrounds the concept of development we could simply conceive it as societal problem-solving, which would imply that a society develops as it succeeds in dealing with predicaments of a structural nature, many of them emerging from the global context. This nonmetaphysical conception of development, however, also implies that efforts at problem-solving may now and then fail, and that most solutions generate new problems – and new tasks for development theory. A major task is to analyse development predicaments stemming from the fact that most decision-makers operate in a national space but react on problems emerging in a global space over which they have only partial and often quite marginal control. This is increasingly true even for so-called superpowers. Hegemonic decline may have its good aspects but it certainly increases the level of uncertainty.

7.3 Three worlds of development

Leaving the rather uncertain future of development studies aside, there are to my mind good reasons for welcoming what has been achieved so far in the process of transcending Eurocentrism and endogenism. Since problems of development vary qualitatively from one society to another, due to both domestic conditions and the position in the world system, it should be the task of development theory to identify specific development problems and find solutions that are appropriate with regard to society, ecology, culture as well as international conjuncture. This specificity of development problems can for our purposes be illustrated by the 'three worlds of development' – at least until the perhaps not so distant time when they are completely dissolved and transformed into some other world-system structure.

7.3.1 The specificity of development

To start with examples from countries in the First World, their 'lead' was not due to any historical law, as suggested by the modernization paradigm, but to a set of concrete historical circumstances, creating advantages of a more or less temporary nature. As Geoffrey Barraclough (1980) puts it: 'The sources of Western predominance dry up'. That the preconditions for predominance are in a process of change will have a deep impact on the power structure within the 'centre'. First of all, the centre itself is moving eastward, thus loosening the historical association between 'capitalism' and 'the West'. Secondly, according to some observers, the new political alignments and conflicts indicate the disintegration of the West itself. Thus, rather than being a temporary recession, the economic crisis in the West to many countries signalled something more fundamental: a development predicament, including problems such as marginalization, deindustrialization and permanent unemployment. This is why development theory appears to be of relevance also in the West.

The Second World faces problems of a partly different nature.

This can be explained by the fact that its development impulse originally was a response to the Western European challenge – the modernization imperative. The Eastern European solution to the development problem was precipitate, marked by specific structural traits, such as excessive centralization, a heavy industry bias, and regional imbalances, which in turn generated other development problems, of which agricultural stagnation, a technological lag and legitimacy deficit perhaps are the most prominent. Of course these problems are aggravated by the fact that the Soviet model was imposed on countries with very different preconditions. Thus the search for more indigenous socialist solutions has previously implied an act of disloyalty, leading to a block disintegration similar to the one occurring in the First World. This is no longer necessarily the case, as long as the reforms retain a socialist quality and the initiative rests with the leading force, i.e. the party. Furthermore, there is now an objective necessity behind the reforms, well understood by the educated, middle-aged middle class, which, in terms of class struggle, is the main actor behind the reform drive. However, the political preconditions for implementation differ considerably from one country to another, although they all face a similar predicament.

The Third World, however, is still, and for good reasons, regarded as the core area of development problems, to which the main part of the book accordingly has been devoted. In this context it should only be stressed that the Third World is going through a process of differentiation, i.e. the hierarchical global structure is reproduced on a lower level with new and different development problems as a result. Thus some countries will cease to be underdeveloped. Others will find themselves 'dependent' on new, this time regional, centres. Theorists who hold on to the centre-periphery model speak about 'semiperipheral' or 'subimperialist' states, but in fact this is not a very homogeneous category. Some of the 'new influentials' – if we confine ourselves to the material aspects – base their status on strategic natural resources, others on successful experiments with industrialization, while a third category combines one or both of these criteria with geopolitical power in the context of their own region (see Fig. 7.3). Any future industrial superpower is likely to emerge in the shaded area of the figure, but are there any serious candidates?

This differentiation of the Third World is also reflected in the efforts to identify new problem categories in the periphery, such as the Fourth World, the Least-Developed Countries, the Most Seriously Affected Countries, the Land-locked Developing Countries, the Island Developing Countries, etc.

The concept of the Fourth World has varying meanings, but the others mentioned have been operationally defined for purposes of classification within the UN system, and constitute distinct,

Fig. 7.3 Differentiation of the Third World: the semiperiphery

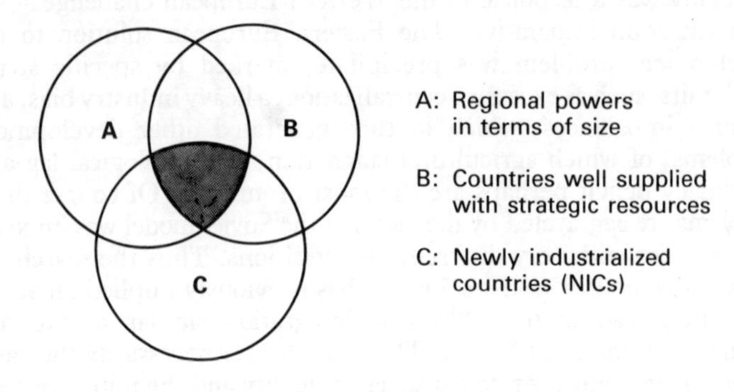

although partly overlapping categories (see Figure 7.4). In this case the shaded areas of the figure indicate a particularly problematic situation.

The concepts are examples of the growing awareness that the Third World is disappearing but do not amount to a (much needed) systematic categorization of the developing world. The hierarchy within the South is indeed in search of theory (Korany 1986). My main point is that the concept of the Third World hardly will survive for long. Hence the title of this book signifies merely a phase in the development of development theory.[107] The time has passed when development theory was believed to be applicable only to a special category of countries. Thus the differentiation supports the theoretical universalization, but this, on the other hand, must be clearly distinguished from a retreat to the false Eurocentric universalism.

Fig. 7.4 Differentiation of the Third World: the new peripheries

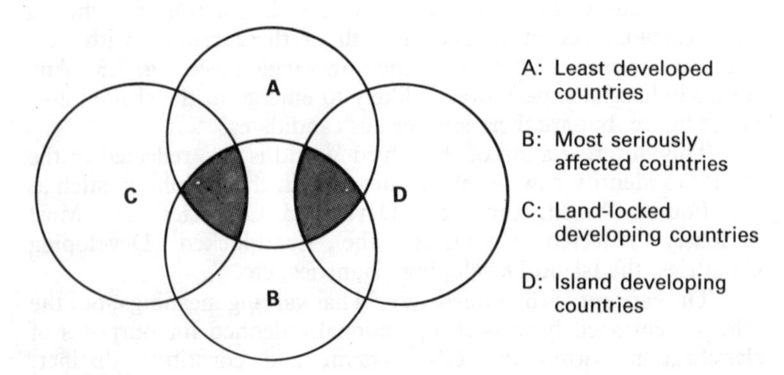

7.3.2 The future of development

The fact that the 'three worlds' are disintegrating and development is becoming a global and universal problem makes it probably too important to be left to a special discipline with low academic status and – on top of that – under fire from all sides. Without a 'special case' there is no need for a special discipline. The 'development problem' comes closer as world space and national space interweave. It will therefore be a concern for all the social sciences and draw them to each other, if not merge them into one unified social science.

Development is a concept which so far has been strongly associated with the state, and with national development strategies. What is the future of development in a world where most states lack the capacity to influence their own economic development? What is the function of national development strategies in such a world?

This raises the issue of the state which – in spite of all the controversy surrounding this concept – stands out as one area of converging interests in development theory. We opened this book with a discussion of the crises of the state in the 'three worlds'. What kind of political structures will emerge from these crises? What kind of political actors will take up the challenge of development in the future?

A second area of converging interests is the nature of the international system – or global system, world society, world system . . . The plurality of concepts reveals the lack of theoretical consensus. However, there can be no doubt about the interest for the 'object' as such. Will the world continue to develop towards interdependence, will it disintegrate into an anarchy of economic nationalisms, or will it be transformed into a regionalized system?

To try to answer these questions would be to start writing a new book. To me, the first of these three scenarios does not seem viable, the second is a road to disaster, whereas the third may turn out to be an enduring solution to the global crisis – and the crises for the nation states in the three worlds. My purpose, however, is merely to indicate what kind of research agenda for future development theorists that emerges from our retrospective review; an agenda freed from Eurocentrism and endogenism and with the appropriate mix of positivism and normativism, formalism and substantivism.

It is futile to draw up perfect models for an inherently imperfect world. It is nevertheless possible to identify basic trends and, as it were, to 'assist history' by resisting unviable, utopian projects and supporting realistic utopias, such as making a shift towards sustainable development and safeguarding cultural pluralism. A disintegration of the present hegemonic world order, which is an integral part of the 'global crisis', would make these 'realistic utopias' look less distant.

In a posthegemonic world the concept of region will assume a new importance as a possible mode of organizing the international economic order, and obviously a better alternative than protectionist regression on the level of the nation state. This in my view is a realistic utopia.[108]

The nation state is a (nonrealistic) utopian project in many parts of the world and will merely lead to a proliferation of nonviable ministates: Khalistans and Eelams. The present ethnic upsurge, which is a worldwide phenomenon, cannot be accommodated within a closed nation-state framework. This is evident also from the European experience. The immediate solution to this problem, which will endure beyond the Cold War, necessitates regional cooperation but it will have to be combined with a long-term solution, implying stronger local autonomy and ethnodevelopment. As objective movements going beyond the nation state (but not implying its disappearance), I thus see two different but complementary types:

- One pointing upwards towards the region – 'the new regionalism'.
- One pointing downwards towards the local community – 'the new politics', 'the new social movements', 'third system politics', 'nonparty politics', 'informal politics', 'noninstitutional politics'. (Again the plurality of concepts indicates an emerging phenomenon, about which there is little theoretical consensus.)

What is new in this? By 'new regionalism' I mean political cooperation on the regional level to promote the region as a viable economic, cultural, and ecological unit. This is different from the concept of 'common market', which so far has merely reproduced the hierarchical division of labour on a regional level, and in most cases ultimately disintegrated due to the inherent contradictions. There are many examples; East Africa and the Caribbean, to cite but two.

In the European context the EEC (more than a common market but not quite a community) can be contrasted with the project of creating a European Fortress, a true European region based on a domestic market, a shared culture and a historical identity. The 'territorial' project may seem utopian, but is it really more utopian than creating a 'functional' organization primarily for the European-based transnational capital? Which project suffers from the deepest contradictions?

By 'third system politics', to pick one of the alternative concepts enumerated above, I mean mobilization from below with the purpose of controlling the fundamental conditions of life: basic needs fulfilment, the right to belong, to have an identity. It is increasingly and widely realized that 'old' politics – dealing with power in functional macrosystems – has become a threat to these fundamental conditions. There are today a great many reports from the Third World about this new kind of politics, emerging to replace the retreating state. It is 'another development' put into practice. In the European context 'third system politics' is expressed through the Green Project. Local self-reliance, favoured by the Greens, would in principle be compatible with the Fortress Europe project, but the two rarely meet.

The process of international economic integration has gone so far that autarchy, even on a regional level, is practically out of the question (besides being rather unattractive in itself). However, with regions as basic units, the fundamental goal of 'good old development theory' – self-reliance – could be viable, whereas self-reliance on the country level never was. The Gandhian approach was self-reliance on different societal levels, with priority for the local community. This approach, however, is quite contrary to the current economic logic. It must therefore be a conscious political creation derived from the lessons of history.

Furthermore, viable interregional relations must also be explicitly political, i.e. created on the principle of symmetry, rather than emerging from the principle of comparative advantage, which, in the absence of political constraints, ultimately ends up in a perverted division of labour. This was the message of development studies as expressed by the great pioneers. It was also the message of the dependentistas which followed. The former had a naive belief in the progressiveness of state intervention, the Dirigiste Dogma as one 'counter-revolutionary', Deepak Lal, has named it (Lal 1983). The latter were disillusioned about the bourgeois state but equally naive in their voluntarist belief in immediate revolution, and their view of socialism as something which could be chosen simply because of its theoretical superiority (Lehmann 1986). The reform debate in Eastern Europe is relevant here.

Development is more complicated than its many doctrines, but the tragedy of the present stage of its intellectual history is that the attacks on what we now may call classical development theory – whether in its reformist or in its radical form – rest on even more simplistic assumptions. For that reason I think it has been meaningful and worthwhile to give an account of this tradition from the point of view of the present 'losers'. History is already being rewritten by those who for the time being enjoy the monopoly of definition.

Notes

Chapter 1

1 Earlier versions of this chapter have been presented at the EADI Workshop on Development Theory in Warsaw, 25–7 March 1987, the Third Annual Conference of CASID in Hamilton, Ontario, 4–6 June 1987, and the EADI General Conference, Amsterdam, 1–5 September 1987. I am grateful for criticism, particularly from Attila Agh, Gunder Frank, Herwig Palme, Emanuel de Kadt, Colin Leys, Manfred Bienefeld, Weine Karlsson and Marek Thee.

2 At the British Development Studies Association Conference in September 1983 Robin Cohen stated: 'There is a crisis in development theory because we have no theory of the crisis'; quoted from Faber, M., 'Development Studies in the UK', *EADI Bulletin*, 1984: 2 (special issue on European Development Studies).

3 See the special issue of the EADI Bulletin, referred to in note 2. Actually the rationale for this special issue was a feeling that European development studies were facing difficulties and the threat of marginalization.

4 One unpredicted event which has caused concern not only in Western but also in Soviet social science is the Iranian revolution. On this, see Martha Brill Olcott, 'The Iranian Revolution: A Crisis of Social-Science Theory', in Thompson and Ronen 1986.

5 Transition is a concept indicating qualitative and systemic change, most commonly used in the Marxist tradition. Maurice Godelier recently defined transition as a period 'when modes of producing, thinking and acting individually or collectively reach internal or external limits and begin to crack, weaken and disintegrate' (Godelier 1987: 447). In the Marxist tradition the greatest attention has been paid to transitions to socialism, a problem to which we will return in the next chapter.

6 At the UNU Seminar on 'Europe's Role in Other Regions' Peace and Security', in Austria, May 1985.

7 An alternative with a high degree of systematization would be to make a categorization of developing countries (oil-producing, industrialized, agriculture, etc.). Such attempts, however, tend to be rather incoherent. For an example, see Beaud 1987.

8 From a diplomatic point of view it is of course always a bit sensitive to point out 'sick' countries. On the other hand we need to concretize the crisis (or the different crises) in order to assess the proposed remedies in a systematic way. I believe that a closer look at Great Britain is illuminating from that particular point of view. For the same

reason we shall let Poland illustrate the crisis in Eastern Europe, Ghana the crisis in Africa, Chile the crisis in Latin America and Sri Lanka the crisis in Asia.

9 'The term "actually existing socialism" (realexistierender Sozialismus) was coined to justify the contrast between what Marx and Lenin had said and what had actually developed.' (Bahro 1984)

10 What Food Self-Reliance (FSR) implies has been summarized by Joseph Collins and Frances Moore Lappé (1980). First, FSR requires a control over agricultural resources by local units. At present agricultural production in many countries is diverted towards low nutrition crops for local elites and foreign markets. Secondly, FSR necessitates mass initiative rather than government managed 'development'. Thirdly, trade must be an expression of domestic need rather than foreign demand. Fourthly, the role of agriculture must change from being a means to becoming an end. Fifthly, industry will have to serve agriculture rather than the other way round. Finally, self-reliance requires coordinated social planning, which is not the same as authoritarian rule from the top.

11 It is no coincidence that perhaps the most interesting. Third World literature in the last decade has been focused on the role of the state. This literature is refreshing since most theoretical schools earlier avoided the issue: political development theory dealt with 'political systems' where the state as such was conspicuous by its absence, dependency theory on the whole had little to say about the state, and what orthodox Marxism had to say was of little use. Recent contributions address the problem of the state head on and from various theoretical perspectives, from Marxism to public choice. (See Bates 1981, Clapham 1985, Collier 1979, Goulborne 1979; Hamilton 1982, Hydén 1983; Skocpol 1979, Stepan 1978, Trimberger 1978.)

12 Development policy is an asymmetrical relationship between the state and its subjects which the late Bernard Schaffer called 'labelling' (Wood 1985). 'Government programs transform people into objects – as recipients, applicants, claimants, clients or even participants' (op. cit: 13). Mainstream development theory can be seen as a part of this process by providing the 'labels': categorizations, abstractions, typologies. Labels can be rejected or accepted and internalized, in which case they define a situation which Gramsci called hegemony. The best example I can think of is the system of 'protective discrimination' in India (Galanter 1984). Schumacher's concept 'small is beautiful' (see Chapter 5) is an example of counter-labelling, obviously inspired by 'black is beautiful' and intended to undermine prevalent beliefs in the economies of scale. Counter-labelling is usually in turn countered by ridicule from authorities and guardians of the 'conventional wisdom'.

13 This terminology was suggested by Marc Nerfin (president of the IFDA) in a lecture in Gothenburg, November 1983.

14 In this context I would like to mention Basil Davidson's lecture 'African Nationalism and the Problems of Nation-Building: Reflections on the Past 25 Years', delivered on the occasion of the 25th anniversary of the Nigerian Institute of International Affairs (December 1986). Davidson repeats his point, made several times in his

earlier works, that none in the range of institutional models offered to Africa by the outside world is going to offer viable solutions. What then, are the solutions?

Davidson points at two promising trends. The first one is devolution of power: 'This trend towards mass participation in self-government seems to me to have become a genuine manifestation of Africa's capacity to find and apply its own solutions.'

The second trend is a resurgence of the old ideas or values of African Unity, albeit in very different circumstances: 'Now the peoples of Africa are voting with their feet for some kind of unification, a kind of "peoples" Pan-Africanism.'

Chapter 2

15 In M. Marien's survey *Societal Directions and Alternatives: A Critical Guide to the Literature* (Marien 1976) no less than 350 titles for modern societies were identified.

16 The modernization imperative presents itself in different forms to different countries. The origin of this dynamics, which explains the basic similarity of the modernization process in various parts of the world, was the European arena where the superiority of Great Britain represented a threat to the latecomers. Later, when the modernization process became global, the threat did not necessarily come from the most advanced industrial countries in absolute terms, but rather the more advanced countries in relative terms within particular regions. Tanzania, for instance, perceives the threat as coming from Kenya, Argentina wants to catch up with Brazil, Pakistan with India, etc.

17 In what is bound to become a rival interpretation to that of Bury, Nisbet (1980) argues that the idea of progress in fact goes back to the very beginnings of Western civilization. This may very well be correct. However, the existence of the idea of progress (the belief that mankind has advanced in the past, is now advancing and will inevitably advance in the future) among a handful of philosophers should be distinguished from the rise of the idea to a social paradigm, which undoubtedly is a modern phenomenon. A distinction between the idea and the doctrine of progress might be helpful. Only the latter would then have relevance for understanding the collective behaviour of the Westerners. Nisbet seems to have a similar distinction in mind when he discusses the end of progress:

> When the identity of our century is eventually fixed by historians, not faith but abandonment of faith in the idea of progress will be one of the major attributes. The scepticism regarding Western progress that was once confined to a very small number of intellectuals in the nineteenth century has grown and spread to not merely the large majority of intellectuals in this final quarter of the century, but to many millions of other people in the West. (Nisbet 1980: 317)

18 Bury does not consider the idea of progress in Marxism. For this see Pachter 1974.

19 Many would perhaps object to the inclusion of Marxism in Western development thinking, since adherents to Marxism usually stress its universal applicability. Samir Amin, for example, holds that Marxism was a radical break with the tradition of European intellectual history. He argues that Marxism, although it came into being within the European cultural sphere, 'transcended the debates of that society, thereby acquiring its universal validity' (Amin 1977b: 28)

Of course this is very much a matter of interpretation. Even Amin concedes that what he calls 'the vulgar reduction of Marxism to economism' was in line with the spirit of the time and gave Marxism 'a western-centered character which it shares with the ideology of capitalism'. It is only natural that the most typical expressions of a tradition of thought are not the most sophisticated variants, since sophistication implies a certain awareness of problems of application.

Liberalism and Marxism thus both form part of the mainstream of Western development thinking and are, according to Johan Galtung, 'two ways of being Western'. Galtung feels that it would be strange if liberalism and Marxism should not be more similar than different, since they developed at about the same time, in the same place, and both reflected a particular culture dominated by the capitalist system (Galtung, *Two Ways of Being Western*, n.d.). The missionary assumption (the idea that the West shall save the world), the idea of nature as an object (to be mastered and exploited) and of society as governed by laws are some of the shared biases. Both ideologies, furthermore, stress the primacy of industrial production, as opposed to agricultural production, and look upon the nation state as a basic unit rather than, for example, the local community or, one might add, the 'world system'.

20 The rather extreme degree of Westernness implicit in this imitative approach is discussed by the author in a comparison between Witte and the pioneer of planning in India, M. Visvesvaraya (Hettne 1978).

21 Kurt Martin (Mandelbaum) tells the following revealing story: 'I remember a conversation around 1950 with a Minister in a newly independent Asian country. When I kept on stressing the need for doing more for agriculture he said that it was the imperialists who always told them to concentrate on agriculture; they had thought that I was an industrialization man.' See the special issue of *Development and Change* (p. 511) referred to in note 23.

22 The (self-) presentations together with commentaries have been published in Meier and Seers 1984. A second publication (Meier 1987) dealt with a later period described by the editor as 'resurgence of neoclassical economics' and containing the following pioneers: T. W. Schultz, G. Haberler, H. Myint, A. C. Harberger and C. Furtado. The description is of course not applicable to Furtado who represented the dependency school. The 'resurgence' will be discussed later as 'the counter-revolution'.

23 This interventionist orientation is even more obvious if one takes into consideration pioneers outside the World Bank list, which of course should not be seen as exhaustive. For instance Kurt Mandelbaum as early as 1945 published *The industrialization of backward areas* which

actually was a programme for a planned transformation of an agrarian economy into an industrial one. Mandelbaum's own account of the background to this book can be found in *Development and change*, **10** (4) Oct 1979. In the same issue, devoted to the work of Kurt Mandelbaum (Kurt Martin), Hans Singer reviews the book, definitely a pioneering work.

24 The paper entitled 'The Distribution of Gains between Investing and Borrowing Countries' was printed in *American Economic Review, Papers and Proceedings*, **II** (2) May 1950. Reprinted in Singer, 1978, in which a retrospective commentary, 'The Distribution of Gains Revisited' also can be found.

25 When Myrdal's magnum opus *Asian Drama* (1968) was published, the modernization ideal that it stubbornly adhered to was already severely battered, and its impact was less than expected. Its critical function must, however, not be underrated.

26 This is the subtitle of John Toye's important analysis Dilemmas of Development (Toye 1987a) on which I rely heavily in this section. I am also grateful for his comments and suggestions. The counter-revolution in development economics is obviously part of a larger change in economic thought, termed 'the barbaric counter-revolution' by W. W. Rostow (1984):

> The counter-revolution is endemic in the industrialized North. It is being conducted in some countries reluctantly and defensively. In Great Britain it has been pursued with conviction and verve since the accession of Mrs. Thatcher's conservative government in May 1979. Its center, however, is in the United States. There the counter-revolution can be dated arbitrarily from the imposition of a restrictive money-supply policy by the Federal Reserve on October 6, 1979 . . .

27 'Orthodox' is of course a relative concept, the meaning of which is contextually defined. During the rule of the structuralist paradigm the counter-revolutionaries experienced themselves as unorthodox. Little says correctly about Haberler and Viner: 'If one judges orthodoxy by the criterion that those are orthodox whose views largely coincide with those of the policy-makers, or with the spirit of the times, there is no doubt about their unorthodoxy' (Little 1982: 75). Here, I, taking the longer view, use orthodox as synonymous with neoclassical. Unorthodox would among other positions include the structuralist view. The two schools go by different names. Little argues convincingly for using neoclassical v. structuralist as the main distinction (op. cit: 19).

28 Many of P. T. Bauer's views on development, particularly his dislike of government intervention, go back to earlier studies of colonial marketing boards and their exploitation of peasant producers. See for instance his West African Trade (1954).

29 The 'crisis approach' in political development theory only referred to crises internal to a particular political system. In an attempt to rescue this approach Rothchild and Curry (1978: 95) add two externally generated crises: the survival of the nation as constituted at independence; and securing freedom from external control.

30 That the 'Marxist' conception of five stages as a unilinear process is a misunderstanding of what Marx really meant does not minimize the impact of this idea. Already his mentioning of the Asiatic mode of production destroys the scheme (Hobsbawn 1964, Melotti 1977). The simplistic historicism had to be modified and elaborated by later Marxists in order to maintain the credibility of the Marxist approach. In the first place it could be shown that the popular 'Marxist' view of historical change probably never was Marx's own, and that he himself thought of a multilinear, rather than a unilinear transition to socialism.

31 The variety is so marked that it would be very difficult to reach a consensus about which countries ought to be regarded as socialist. According to Gurley (1978) there are about twenty-four socialist countries, producing a fifth of the world's GNP and having roughly a third of the global population and land area. Fourteen countries are in the Third World. Gurley's list includes some doubtful cases such as Guyana and Benin, but on the other hand there are authors who consider Burma, Iraq and Tanzania (not in Gurley's list) as socialist as well (see the special issue of *World Development on socialist models of development*, **9**, 9–10). Gordon White (see White et al. 1983 and White's article in Kaplinsky 1984) makes a distinction between Marxist-Leninist socialist countries and socialist 'intermediate regimes' with further subdivisions based on degrees of industrialization and level of GNP. The first category contains twenty-four countries, the second fifteen (with some question marks). However, even with a fairly narrow definition of socialism, one is forced to conclude that no single socialist path to development exists today.

32 For a Soviet explanation of the path of noncapitalist development – or as it also came to be called, 'socialist-oriented development' – see Solodovnikov and Bogoslovsky 1975. According to these authors, what noncapitalist development essentially amounts to is that 'in the era of imperialism and socialist revolutions, the peoples of backward countries can achieve socialism not through capitalist development, but bypassing it or interrupting it at an early stage' (op. cit: 24).

Over time the tendency has been to stress that socialist-oriented countries are, and have to remain for long time, parts of the capitalist world economy, and to de-emphasize the role of their links with the Soviet bloc. Also, Soviet scholars have begun to doubt the possibility of bypassing the stage of capitalist development.

Chapter 3

33 To cite but two examples, one from the right and one from the left, P. T. Bauer (1971: 148) asserts that 'it is untrue to say that colonial status is incompatible with material progress', whereas Bill Warren (1980: 129) concludes that 'the colonial era, far from initiating a reinforcing process of underdevelopment, launched almost from its inception a process of development'. It goes without saying that none of these authors recognized any need for indigenization of development theory.

34 This was on the whole confirmed at a meeting of regional social science associations at the SID conference in Delhi, March 1988.

35 Furtado's analysis of the Brazilian economy during the depression can be found in his The Economic Growth of Brazil (Berkeley: University of California Press 1963). The Brazilian edition was published in 1959. He first discussed this in 'Caracteristicas de Economia Brasileira', *Revista Brasileira de Economia* (March 1950). For a retrospective discussion, see Furtado's contribution and the comments by Vittorio Corbo in G. M. Meier (ed), *Pioneers in development*, Second Series, published for the World Bank by Oxford University Press 1987.

For neoclassical economists the policy was wrong but for the dependentistas it was inadequate, i.e. not radical enough. But both groups were equally critical.

36 'The State-as-planner as characterized by ECLAC was seen as undertaking a greater degree of intervention than that suggested by the Keynesian model because its objectives were not confined to guaranteeing full employment, stimulating economic growth (how much to produce) and ensuring a more equitable distribution of income (how to distribute), but also included restructuring the economy for industrialization (what to produce), orienting the utilization of the factors of production in accordance with their availability (how to produce), and accomplishing all this from a peripheral position which made it necessary to exercise a much firmer control over external economic relations and their impacts . . . However, ECLAC did not ask itself, at the time, whether the Latin American states were up to these tasks, if they could carry them out successfully.' (Gurrieri 1987)

37 For a general summary of different national experiences, see special issue of *World Development* (1977: 1–2) on 'Latin America in the Post-Import Substitution Era'. Cf. also the discussion of development strategies in the following chapter.

38 It should be admitted that a brief account may not correspond to the way this intellectual process actually was experienced by the founding fathers of dependencia. Fernando Henrique Cardoso, for example, points to the difficulty of giving a true description of the intellectual prehistory of a new paradigm. He may be right in asserting that the marriage between neo-Marxism and the CEPAL approach is a simplistic way to explain its rise, which certainly was a more complex and composite intellectual process (Cardoso 1976). On the other hand it is also important to understand how a certain stream of ideas was conceived by others than the actual carriers of these ideas. This makes some simplification both necessary and legitimate.

39 Some Latin American classics within the dependencia tradition are: F. H. Cardoso and E. Faletto (1969); O. Sunkel and P. Paz (1970); Rudy Mauro Marini (1973) and Vania Bambirra (1974). This brief survey draws heavily on a more comprehensive study based on interviews with many of the dependentistas. See Blomström and Hettne 1984.

40 Interview with Pedro Paz (UNAM, Mexico City, December 1977).

41 Here I would like to refer to an interview with Owen Jefferson (Kingston, November 1978), one of the dependency theorists in

Jamaica who under Manley's regime worked as deputy governor of the Bank of Jamaica. Asked whether he had changed his theoretical position in view of Jamaica's discouraging economic performance he replied:

> I don't think it has changed my view of the way the system actually operates with regard to the dependency syndrome but what it has certainly impressed on me is the difficulty of breaking out of it. . . . One of our assumptions was that ownership of resources does in fact constitute control, which is not really true because of the very important technology factor and the fact that the market is very unpredictable.

42 Interview with Pedro Vuscovic (Mexico City, Oct 1981).
43 Statement published in *Monthly Review* (1981: 2).
44 Rather, the conflict is internalized in the cleavage between Ladinos and Indios.
45 For a discussion of the polarity of Mexican thought which also makes use of the concept of 'counterpoint', see Weinstein 1976.
46 Mariátegui's homespun, nationalistic Marxism (somewhat reminiscent of Mao Zedong) was consistent with the Comintern line until the reins were tightened in 1928; after that he turned into an eccentric in an increasingly streamlined communist movement.
47 On this Kari Polanyi-Levitt says: 'The germ of "Plantation Economy" was born in 1964 in St. Augustine in a very exciting series of all-night discussions with Lloyd Best and Alister McIntyre. I remember we taped some of it (for posterity!) but the tape was burned in fire in Lloyd's Tapia House a few years ago.' (Personal communication 1988.)
48 There was a strong connection between Caribbean and Canadian Social Science, since many Caribbeans studied at Canadian Universities and at least one Canadian, Kari Polanyi-Levitt, had a long association with West Indian Universities (cf. note 47 above). Kari Levitt founded a 'New World Group' in Montreal and applied the dependency approach to Canada in a 1968 article in New World Quarterly, later expanded into a critique of Canadian dependence in Silent Surrender (Levitt 1970).
49 Interview with George Beckford (Kingston, Dec 1979).
50 This was a most explosive analysis under the circumstances and the propagation of the message forced him into exile. Ultimately, Rodney was murdered in his home country Guyana.
51 This argument does not apply to the smaller Asian countries which, however, cannot be covered in this context. The smaller countries surrounding India have their own dependence syndrome: the Indian giant. Burma was a case of nationalist self-reliance recently breaking down in a national upsurge against imposed isolation. The communist countries in South East Asia were in varying degrees influenced by the Maoist doctrine which emerged as an alternative to the Soviet model, Kampuchea being the obvious case in point. Vietnamese social science was on a more orthodox line, the overall task being 'to make use of Marxism-Leninism and of dialectical and historical materialism in order to carry out fundamental problems of a

new Vietnam' (Dahlström 1988). The capitalist part of South East Asia, i.e. ASEAN, is following a different road, to be discussed later, but in a country lika Malaysia we, not unexpectedly, find a minority view stressing the problems of dependence and the need for self-reliance (Khor Kok Peng 1983).

52 Interview with Rajni Kothari (Delhi, April 1976).

53 It should, however, be noted that one of the major figures in early Indian Marxism, M. N. Roy, took a position that deviated from mainstream Marxism and anticipated the dependency perspective:

> The introduction of higher means of production in cotton manufacture marked an era of social and economic progress in England, but it had retrograde effect upon India. The forces that helped to build so many industrial centres in the former, were used for destruction of prosperous towns and urban industrial centres in the latter. The reason of this diametrically opposite effect of the same cause was that India as well as English society came under a more developed method of exploitation, but the improved means of production which made this new method of exploitation possible, remained the property of the bourgeoisie of one country, which became the political ruler of the other. (Roy 1971 [1921])

54 In fact the journal was launched on this very issue.

55 At one such occasion it was stated:

> Gandhi's ideas in the fields of Politics, Economics, Sociology, and even in Anthropology seem to have emerged from the creative thinking of a genius perhaps unparalleled in the history of mankind. We have legions of social scientists these days with their pretentious platitudes and slogans, but Gandhi's contribution to social sciences bears the distinct stamp of his personality. If we feel desperate these days in the matter of our social problems, it is because we have not so far tried to give a practical shape to the ideas of one who in the manner of a scientist, carried on his experiments on truth. (Vidyarthi 1970: ix)

56 A brief treatment may be found in B. Hettne and G. Tamm 'The development strategy of Gandhian economics', Journal of the Indian Anthropological Society, 1971: 1.

57 A typical example of how Nyerere conceived of socialism would be:

> Both the 'rich' and the 'poor' individual were completely secure in African society. Natural catastrophe brought famine, but it brought famine to everybody – 'poor' or 'rich'. Nobody starved, either of food or of human dignity, because he lacked personal wealth. He could depend on the wealth possessed by the community of which he was a member. That was socialism. That is socialism. (Nyerere 1968: 3–4)

I am here bypassing the many different interpretations of Tanzania's experiment in socialism. See Issa G. Shivji's Class struggle in Tanzania (London: Heineman, 1976) and the discussion in Review of African Political Economy, Nos 3 and 4 (1975).

58 Tony Killick makes the following assessment of the development strategy carried out by Nkrumah:

> In effect Nkrumah mistook the appearances of modernization for the reality. Much effort and many resources were devoted to acquiring the symbols of modernity – the institutions, the machinery, the factories, the apparatus of state. The results were modernization without growth,

occuring, I submit, because the strategy overlooked that past industrial revolutions have drawn their dynamism from innovation and adaptation. There was little of either in the many changes introduced by Nkrumah. Most 'innovations' were copied unchanged from some Western or Eastern model. They worked out differently, but the differences were not to Ghana's advantage. (Killick 1978: 336)

59 Interviews with P. Mbithi and K. Kinyanjui (IDS) in Gothenburg 1977.

60 Thandika Mkandawire (1988) from CODESRIA has an example:

> The worst case of this sort of aloofishness was the 'Kenya Debate' conducted among a few expatriate scholars on what was the nature of the state in Kenya. By a process of self-reinforcing cross-referencing among themselves, these scholars were able to create a veritable enclave of intellectual discourse from which native scholars were virtually excluded. With the end of their contracts and their return to their respective homes the 'Kenya Debate' ended.

I owe the title of this subsection to Thandika Mkandawire, who also points out the ironic fact that:

> When most African governments insisted on their national priorities there were few indigenous social scientists and most of the experts were expatriates who were not bound by national priorities. Now that Africa has large numbers of social scientists, African governments have significantly lost degrees of autonomy and in one way or another are pursuing objectives imposed by external financial institutions. Thus two or three IMF experts sitting in a country's reserve bank have more to say about the direction of national policy than, say the national association of economists.

61 Interview with Samir Amin (Gothenburg, March 1981).

62 Abibirim is an Akan word with connotations of Blackness or Africanity. The philosophy behind the Abibirim strategy is based on the following premises (Kofi 1974: 32):

- We cannot formulate adequate economic development theory for the indigenous African economies without a careful analysis of the role we have placed and continue to play in the development of Western capitalism, and also peripherally in the socialist camp.
- For a viable economic development we must change our internal structures and external relations. Internally, our sociocultural milieu prevents our societies from utilizing effectively the surplus we generate to spur development. Externally, the dependent role we play in the world capitalist order inhibits our growth.
- The African countries should examine critically the Western capitalist economic theories, since they were developed for a different social order, and modify them, if need be, before using them.
- Traditional African countries should critically study and examine the socialist models, adapt them and use the more relevant aspects.
- African states must work closely together to develop and enforce development strategies and learn from each other. African unity will only make sense if a common ground is established for a development policy.

63 In Mexico under the presidentship of Luis Echeverria, the dependency perspective was even incorporated into school textbooks (Cockroft and Gandy 1981).

Chapter 4

64 The OECD list includes ten countries: Singapore, South Korea, Hongkong, Taiwan, Brazil, Mexico, Spain, Portugal, Yugoslavia and Greece. The World Bank list adds Malaysia, Argentina, Turkey, the Philippines, Columbia and South Africa. (See Hoogvelt 1982: 25)

65 Apart from Fernand Braudel, who has given his name to the centre for world-system analyses at Binghamton, New York, influential Annales figures in this context have been Marc Bloch and Pierre Chaunu. It should be pointed out that world-system theorists differ from the annales researchers in that they make little use of primary sources. They use history rather than write it.

66 The concept 'external arena' has a residual quality about it which tends to simplify both the vast differences between different political and socioeconomic systems in the extra-European world, and the variety of processes, mechanisms, conflicts and alliances which marked their incorporation. For an attempt to analyse this process from the point of view of the 'people without history', see Wolf 1982.

67 Explicitly, this thesis was adopted by the Sixth Congress of the Communist International in 1928 on 'The Revolutionary Movement in the Colonies and Semi-Colonies'.

68 Robert Brenner, an influential critic of both dependency theory and the world-system approach, stated:

> The appearance of systematic barriers to economic advance in the course of capitalist expansion – the 'development of under-development' – has posed difficult problems for Marxist theory. There has arisen, in response, a strong tendency sharply to revise Marx's conceptions regarding economic development. In part, this has been a healthy reaction to the Marx of the Manifesto, who envisioned a more or less direct and inevitable process of capitalist social productive relations and, on this basis, setting off a process of capital accumulation and economic development more or less following the pattern of the original homelands of capitalism. (Brenner 1977: 25)

69 Comparisons have been made between the Western discussion on the articulation of modes of production and the Soviet concept of multisectorality in developing countries. This was an offshoot of the rediscovery of Marx's precapitalist social formations. See A. Sulejewicz, 'Coexistence and the Modes of Production', in EADI Bulletin, No. 2, 1986.

70 This was early recognized by unsympathetic observers, for example Wall Street Journal (17 July 1975):

> In truth, this new cold war is not really about economics at all, but about politics. What the Third World is saying is not that it needs our help but that their poverty is the fault of our capitalism.
>
> (Sauvant and Hasenpflug 1977: 65)

71 For a collection of representative responses, see Towards One World, edited by the Friedrich Ebert Foundation.

Chapter 5

72 As Gavin Kitching points out (Kitching 1982: 15) 'progress' was a nineteenth century word for 'development' and the debate about progress in those days has direct parallels in contemporary debates in development studies.

73 This terminology was suggested by Marc Nerfin (president of IFDA) in a lecture in Gothenburg, November 1983. What follows is much inspired by that lecture.

74 Such a strategy is developed in *What Now, The Dag Hammarskjöld Report*: 63–87.

75 For a historical case study of the conflict between the emancipation of the state and mobilization of the people in the context of indirect colonial rule, see Hettne 1978.

76 More recently this issue has been discussed again in Redclift 1987. There is no agreement whatsoever on the question whether it is possible to consider the environment within the governing economic paradigm:

> Economists are interested in scarcity as the underlying reality behind human choice. Environmentalists are concerned that economic growth is the reality which makes human choice less and less possible under conditions of scarcity. (op. cit: 38)

77 Costis Hadjimichalis fears that approaches such as 'development from below' and 'New Territorialism' are likely to be trapped in another cycle of good intentions and unmet expectations, paralleling the experience of the postwar planning theories they criticize (Hadjimichalis 1987). He gives an example in the spatial policies of the socialist government in Greece in the period 1981–85:

> The rhetoric about decentralization and people's participation not only was contrasted with the hard reality of selfish Greek capitalism but soon fell apart due to contradictory government policies (ibid: 26)

According to Soja there is a risk of idealizing, or even 'fetishizing', the territorial community in the new approach (Soja 1983).

78 These ideas draw on a lecture by C. A. O. Van Nieuwenhuijze at an EADI seminar on culture and development, Antwerpen, Oct. 1987.

Chapter 6

79 'I believe it was the correct thing to say at that time because not only were growth models and other theoretical devices derived from Western experience, but they were also imported into the universities, especially in Asia and Africa. Today one would say they were part of "cultural dependence". Afterwards I became increasingly sceptical about the relevance of these concepts for European countries as well. When I went to work in Portugal I felt I was in a Latin American situation: rural poverty, dependence on the transnationals, the type of bureaucracy, the educational systems, etc. I felt that some of the theories developed in Third World countries, for example

dependencia might be relevant there as well. And if Portugal why not Greece, Spain, Turkey, etc. In these cases I found that there were some elements in common with what I have observed in Third World countries. In a sense this is contrary to what I said in the 1963 article. The connecting link is that in that article I was saying that neoclassical, and I would add, crude Marxist theories should not be imported into Third World countries where they did not fit but had to be replaced by different approaches. Now I would say, perhaps those theories did not fit the First World either?'

(Interview with Dudley Seers in Brighton, June 1980)

80 'The contrast between the liberal, and the no-growth position is exemplified by a comparison of The Limits to Growth volume, sponsored by the Club of Rome, with A Blueprint for Survival, issued at virtually the same time in 1972 by a British group associated with The Ecologist magazine. The former advocated a state of global equilibrium shifting economic preferences of society toward more services, while the latter attacks the industrial way of life itself and advocates decentralization of polity and economy at all levels.'

(Marien 1977: 422)

'The principal defect of the industrial way of life and its ethos of expansion is that it is not sustainable.'

('A blueprint for survival', *Ecologist* **2**(1) 1972: 2)

81 There are very few studies on what happens with the quality of life in those lifestyle experiments carried out so far. A Danish study suggests that the level of welfare is lower, but the level of security, identity and freedom higher in these communities than for the ordinary Danish population. However, such comparisons would probably give varying results from time to time and, to be comparable, they should also take into consideration the various class backgrounds of the participants. See K. Lemberg *et al.*, *Dominant ways of life in Denmark/alternative ways of life in Denmark*. United Nations University, GPID Project, Tokyo 1980.

82 I have dealt with these seven European social projects in Hettne 1986 and 1988b. In this context I am first of all stressing the alternative projects, secondly I am more concerned with the development dimension than with the security dimension of the projects.

83 Brandt himself later referred to the meeting as 'a modest, partial success. At least one could learn from it'. (Brandt 1986: 108)

84 Thus this quotation from *Business Week* (9 July 1979) is rather a description of revived mercantilism than neomercantilism:

> Three decades of open, free trade that permitted the multinationals to blossom are giving way to a period of neomercantilism . . . the cooperative effort to create an interdependent world . . . is being replaced by what appears to be a free-for-all among industrial nations to grab or preserve as much as possible for themselves of the shrinking pie.

85 A different distinction between populism and neopopulism is made by Gavin Kitching:

> 'Neopopulism' is distinguished from populism in that it is not a purely anticapitalist doctrine, but rather opposes all forms of large-scale

> industrialization including state socialism. It is also distinguished from populism in the far greater sophistication of its economic arguments, and its willingness to challenge industrialization strategies directly, on the basis of their own economic rationale. (Kitching 1982: 42)

To my mind these two criteria are not sufficient for making the distinction. Regarding the first, one would assume that for the populists the organizational form of the industrialization was of secondary importance in comparison with the threat to community it represented. Regarding the second a more sophisticated articulation is simply a change in the form of expression towards a 'scientification' of the ideology, rather than a change in ideology as such.

86 These two areas are here called 'Eastern Europe', whereas both of them, together with nonaligned Yugoslavia and Albania, are referred to as 'Socialist Europe'. However, depending on the context, Eastern Europe sometimes means Eastern Europe with the exclusion of the Soviet Union.

87 The standard work on the Soviet industrialization debate is still Erlich, 1960 (second edition 1967). Moshe Lewin (1975) links this debate with the debates of the 1950s and 1960s. Bukharin's 'Notes of an Economist' (first published in Pravda, Sept 1928) has been published in Economy and Society 8(4) 1979 with an introduction by Keith Smith. An authoritative statement is found in Preobrazhensky, 1965 with an introduction by Alec Nove.

88 One of the more fascinating links was G. Feldman who represented the 'teleological' as against the 'genetic' approach to planning (Carr and Davies 1974: 839). This approach was an elaboration of Preobrazhensky's industrialization strategy, and inspired Evsei Domar in his work on what was to be known as the Harrod-Domar model (discussed in Chapter 2), as well as the pioneering architect of the Indian second plan, P. C. Mahalanobis. The latter went to Moscow in the early 1950s to consult Soviet planning experts and subsequently received Soviet statisticians at his Indian Statistical Institute at Calcutta (Clarkson 1978).

89 On the potential flexibility of the Soviet Model let me quote Moshe Lewin:

> The party monopoly of political power, its declared aim of the reconstruction, socialization, and nationalization of some key sectors in national economy were, to be sure, constant throughout the existence of the Soviet state. Otherwise, such essential features of socioeconomic systems as the scope of the state sector, the relations between it and private sectors or those between the state and social classes, the scope for admitted or tolerated market and monetary mechanisms, the character of economic and noneconomic incentives and their interrelation, and finally, the degree of cultural plurality, as well as scholarly and even political debate outside and inside the party were variables which combined into different patterns or 'models'. (Lewin 1975: 73)

90 It must be kept in mind that no transition to socialism in the original conception as a more advanced stage beyond capitalism has so far occurred anywhere in the world. In practice socialism has been a

model for the latecomers wanting to 'catch up'. Many of these countries had by that stage already had frustrating experiences with capitalism.

The term 'actually existing socialism' was made popular through Rudolf Bahro's *The Alternative* (Die Alternative. Zur Kritik des real existierenden Socialismus, Köln: Europeisches Verlagsanstalt). Bahro himself tells that the decision to write the book was taken as a response to the events in Prague in August 1968 (Bahro 1984).

91 This total concentration on the appropriate state-market mix puts this debate squarely within the Mainstream tradition. Counterpoint socialism is much less visible. For an argument in favour of 'village communism', see Bidelux 1985. In view of the coming environmental catastrophes in several areas a Green perspective on development is bound to grow strong in Eastern Europe as well.

92 In Slovenia there is a number of new social movements similar to those in Western Europe. Obviously there is a strong influence from Austria in operation here. The Slovene communist party now rules in consultation with these reformist groups. This can be seen as a sign of strengthening of Slovenian identity against the rest of Yugoslavia and a pull towards Europe (i.e. Austria and Italy). See the report in *Eastern Europe Newsletter*, **I**, 9(30 Sept. 1987).

93 This issue, i.e. how to combine planned development with democracy, is of course of great relevance for Western socialists as well. According to New Left Review it is 'the most urgent of all questions that confront contemporary socialism' (*NLR*, No. 159, 1986). The problem is, however, how relevant the East European experience is. Should the future of socialism – the feasibility of socialism – be judged from these experiences, as Alex Nove (1983) argues, or should socialists devote more analysis to the potential socialism emerging from the contradictions of 'late capitalism', as Ernest Mandel (1986) argues?

Our earlier analysis has shown that the prospects for building socialism in Eastern Europe were weak to start with. Therefore socialists in the West would do well to analyse the socialist option in the specific Western context. Similarly East European economists seem to have a very romantic conception of societies dominated by the market and tend to disregard earlier frustrating experiments with capitalism in the East. For a rather disillusioned analysis of the Eastern and Western 'lefts' and their aborted dialogues, see Fehér and Heller 1987.

94 In an article in the Soviet party journal Kommunist it was stated: 'The principles of socialism are not principles of charity that guarantee everyone a job regardless of his ability to work' (quoted from *The Economist*, 26 Dec. 1987).

Chapter 7

95 I am not completely alone in this view. Keith Griffin for instance recently stated:

> One is struck in retrospect by the intellectual excitement of the enterprise, by the quality of the contributions of the best and most imaginative thinkers, and by the sense of high purpose of those who tried to grapple with difficult and important problems.
>
> (Plenary contribution at SID 19th World Conference, March 1988, New Delhi)

The emerging consensus outside the field is less generous, however. One of the pioneers, Kurt Martin (Mandelbaum) explains the impasse in terms of a changing agenda: From long term development problems to short run adjustment issues. He does not think that classical development studies (structuralist oriented development economics mainly) has been permanently superseded (Interview, London, June 1988).

96 In this context the discussion (in *World Development*, **16**, 6, 1988) on how to go beyond the impasse, reacting to Booth's (1985) article, should be noted. Let me also mention a few examples of the synthesizing genre. Streeten (1983) discusses a number of development dichotomies and tries to find points of intersection indicating the future course of development economics — 'should it turn out that the news of its death has been exaggerated'. For instance, Streeten speaks of a transition from the 'economics of a special case' to a new global economics of shared problems, but with greater differentiation of approaches and analyses which both unifies and differentiates the subject (op. cit: 21).

Since modernization theory and dependency theory often are presented as antipodes some of the synthesizing efforts try to bridge the gap between them. Apter (1987: 47) thinks that both can be correct simultaneously: 'While each type of theory commits terrible crimes of omission, each not only has something to offer in a renovated "developmental" theory, but is essential to the other.'

Bossert (1987: 319), with reference to the Latin American debate, sees a 'convergence of thought that might encourage the synthetic merging of the perspectives we see evolving in Latin American development theory.'

Finally, Higgot (1983: 103) urges the radical political economist 'to resist the temptation to throw out the modernization baby with the bath water.'

It might be somewhat premature to try to indicate the most viable convergencies but it is important to emphasize that these attempts are being made at the same time as this or that particular contribution is vehemently denounced and the whole theoretical project, as was discussed in Chapter 1, is in grave crisis. Two important convergencies in terms of research problems (with little theoretical consensus however) are the interest for the nature of the state, and the world system, as objects for study in their own terms.

97 Aidan Foster-Carter (1985: 1) notes that 'the sociology of development includes more than sociology alone. It has to be interdisciplinary. You can't just leave out (say) politics or economics.'

On development geography Forbes (1984: 25) says that 'development economics has been the father of development geography, while sociological theory of social change has been the mother'.

Neither does politics in the Third World seem to be a subject in itself: 'The study of politics in the Third World has been closely bound up with changing approaches to the way the Third World is seen as a whole. Political development theory grew out of theories of social change and to a lesser extent of economic development.' (Randall and Theobald 1985: 7.)

Quotations like this could be multiplied. A quick survey renders two observations possible. Economics and sociology seem to be more central, and economists are the ones who make fewest references to other disciplines (although they occasionally refer to something called 'non-economic factors').

98 The positive mode of thought is, as the history of economic man shows, not value-free. It might be argued that the positive trend may be allowed to go too far and end up in positivism, i.e. what is (or seems to be) is also what ought to be. Instead of an approach to the study of social phenomena we are left with an ideology. It is but a short step from stating that man tends to be motivated by material interests to stating that he ought to be so motivated.

99 Let me give two examples. In a widely used textbook Todaro states:

> . . . at least three basic components or core values should serve as a conceptual basis and practical guideline for understanding the 'inner' meaning of development. These core values are life- sustenance, self-esteem and freedom, representing common goals sought by all individuals and societies. (Todaro 1981: 70)

Galtung (1985) similarly describes development theory as a 'holistic approach to human society'. This holism covers four spaces: nature, human, social and world spaces.

100 F. Nixson warns that uncontrolled normativism is inherent in this distinction:

> The recognition that economic growth did not automatically lead to the wider, normatively defined goals of development was thus an important step in the evolution of development studies in the post-war period. Having served its purpose, however, the utopian concept of development is an obstacle to the further evolution of development studies.'
>
> (Nixson n.d: 12)

However, the distinction between growth and development is not a purely theoretical or conceptual problem. It happens to be the case that people interested in growth are not necessarily the people who are interested in development. There is no reason why they should agree on a definition.

101 There is a connection between positive and formal approaches which should be noted. Hirschman makes this point in discussing Adam

Smith's contribution to the 'interest paradigm', stating that the general welfare is best served by letting each member of society pursue his own self-interest:

> While it was a splendid generalization, it represented a considerable narrowing of the field of inquiry over which social thought had ranged freely up to then and thus permitted intellectual specialization and professionalization. (Hirschman 1978: 112)

Thus, the formal approach in social science is related to the emergence of the positive mode of thought, since the latter made scientific inquiry more orderly and less subjectivistic. However, there is no necessary logic relationship between the two, since a formal approach, once it has been established, could be combined with positive as well as normative modes of thought.

102 Seers came back to the theme in his 1979 essay 'The Birth, Life and Death of Development Economics' (Seers 1979), where he stated: 'The logical future today is the study and teaching of development in a social and political as well as economic sense, with a wider geographical coverage and special emphasis on European needs.' Seers saw development economics as 'a transitional stage in the metamorphosis' (op. cit. 717).

More recently Hans Singer (at a panel discussion in SID World Conference, Delhi 1988) stated: 'To a development economist living in an "advanced" industrial country like the UK, the extent to which ideas, concepts and even the language of development economics has filtered into the discussion of our own problems in the "North" is striking indeed.'

As I see it this is true for the whole field of development studies. The project of creating an autonomous discipline seems less and less viable. As Preston (1985: 156) puts it: 'Failure was bult into the project design.' Neverheless it was important in creating space and legitimacy for 'development research'. I am not sure that the time to give up this space has come yet. Rather it is important to defend it against the rising wave of monodisciplinary fundamentalism – or formalism.

103 A normative-substantive orientation has for a long time been the programme of the institutionalist school which can be seen as a North American heterodox 'sect' in economics. For a comparison between the institutionalist theory of economic development and Latin American structuralism, see Street 1987. Of relevance in this context is Street's observation that dependency and institutionalist theory are both holistic, that the common approach is interdisciplinary, and that they make explicit normative judgements, 'in contrast with economists who still assert a preference for positivism and eschew explicit value judgements, while at the same time implicitly upholding the accepted values of a market system' (op. cit: 1880).

104 This theme can be illustrated by the process of indigenization/universalization, changing the content of the Marxist tradition. In Chapter 3 we analysed the rise of neo-Marxism as a response to the biases and weaknesses in Marxist studies on underdevelopment. Just like other Western theories, Marxism had been rather mechanically

applied to the underdeveloped countries. This was bound to give rise to indigenous responses. Thus, Maoism has been interpreted as Marxism in a Chinese context and many dependentistas similarly thought of dependencia as Marxism applied to Latin American reality. Whether Maoism and dependencia should be considered to be within or outside Marxism is to a very great extent a matter of interpretation.

The relationship between Marxism and Maoism is ambiguous. Populism was an integral component of the Maoist version of Marxism. One of the fundamental characteristics of Maoism was the tension between Leninist-type elitism and the belief in the spontaneous socialist consciousness of the peasantry (Meisner 1979). Thus, Maoism could be described either as a sinification of Marxism, or as Marxism perverted by a parochial impulse. Many Marxists, however, would argue that the Maoist concept of 'cultural revolution', stressing that radical changes in the economic base did not necessarily lead to a more socialist superstructure, was a contribution of universal relevance.

The line between Marxism and neo-Marxism prevalent in Latin America is also problematic. An orthodox Marxist would claim that the neo-Marxists have either misunderstood basic Marxist ideas (i.e. they are poor Marxists), or that they should not be thought of as Marxists at all. However, neo-Marxism could be used as a heuristic concept to indicate a stage in the history of Marxist theory with regard to its ability to analyse the problems of underdevelopment. The neo-Marxists have accepted the changes through which Marxism passed after its transplantation from European to non-European soil, while the orthodox Marxists are more concerned about the purity of Marxism. Contrary to Foster-Carter (1974) I believe that it is inadequate to speak of two distinct schools; instead, we are witnessing a dialectic, intellectual process, which can be seen as a phase in the universalization of the Marxist tradition. The theoretical deficiencies in Marxism, with regard to its analysis of the actual conditions in the Third World, are now being mended (some attempts were discussed in Chapter 4).

The analysis of Third World development may also have an impact upon the very core of Marxism. Samir Amin's early works on African underdevelopment can, in line with our discussion above, be seen as Africanized Marxism. One of his later books, *Class and Nation* (Amin 1980), however, goes a step further and provides an example of universalization. Amin rethinks basic concepts in the tradition of historical materialism and asserts that the classic line of development (slavery–feudalism–capitalism) is mythical. The much discussed contrast between a 'European' and 'Asian' development path merely expresses Eurocentrism. What Amin calls 'the unity of universal history' is, however, recreated by the necessary succession of three families of mode of production: communal modes, tributary modes and the capitalist mode.

105 As the reader will recall, Hegel once affirmed that a world spirit unfolded itself through a series of national spirits.

106 The interesting question is not, in Streeten's formulation: 'do the

developing countries benefit or lose from their coexistence with developed countries?', but 'how can they pursue selective policies that permit them to derive the benefits of the positive forces, without simultaneously exposing themselves to the harm of the detrimental forces?' (Streeten, 1983).

107 According to Nigel Harris (1987) 'the end of the Third World' is already at hand, and from this follows the end of 'Third Worldism', i.e. the body of theories which make up the core of development studies. As the present book shows I find the judgement premature, simply because I define the subject area (if not the discipline) of development studies in terms of problems rather than countries. In a changing world it makes no sense to develop a theory for a special and fixed category of countries.

For the same reason I believe that the concept of 'Third World' will have to go. From the very start it suffered from the disease of 'overhomogenization' (Korany 1986). The debate on the origin of this concept was started in the inaugural issue of Third World Quarterly (January 1979) and was returned to again in the October issue 1987. Leaving aside the question who first coined the term, there is a general agreement that the concept implied a third – non-aligned – force between the superpowers and the two dominant socioeconomic models: capitalism and socialism. Thus it was a completely normative concept. Nonalignment is no more a prominent feature of these countries, which is one of the aspects of dif- ferentiation and disintegration – hence the end of the 'Third World'. I do not see the attempts to develop a new category – the 'Fourth World', meaning the least developed countries – as very useful. This is a statistical category which may have its bureaucratic uses but provides no help in our attempts at mapping the world. Of greater significance is the concept of a 'Fourth World', with reference to aboriginal peoples, which exists as nonassimilated islands in the sea of nation states. It is strikingly similar to the original meaning of the Third World: a heterogeneous collection of units sharing an objective common interest, in this case cultural survival.

108 This is different from the classical Listian argument for a coherent national economy which is not a viable proposition today. John Maynard Keynes still thought in terms of 'national self-sufficiency' when he in 1933 questioned the value of free trade for peace, a remarkable thing to do for an Englishman, as he himself also emphasized (Keynes 1933: 755). Keynes saw a certain degree of national self-sufficiency as a precondition for peace. His discussion of what today is called Self-Reliance focused upon the options of the single state and this – as the 'neo-Keynesians' have discovered – is the limitation, particularly in the integrated world economy of today.

In an interesting article from 1945 Karl Polanyi developed a regionalist scenario against attempts to reshape the liberal world order. This article was written with the hindsight of the Second World War but it is nevertheless interesting to compare with Keynes's 1983 article. Both were primarily concerned with the question what kind of international economic structure was most conducive to peace. Both of them warned against a liberal world order, but while Keynes emphasized the need for national self-sufficiency, Polanyi saw the solution in regionalism (Polanyi 1945).

The 'regions' discussed by Polanyi were the British Commonwealth and the Soviet Union. It is interesting that he could conceive of a regionalist scenario on such weak foundations. Obviously he underestimated the United States, and overestimated England and the Commonwealth, which was left to disintegrate. Instead England joined the US in creating what we have earlier discussed as the Atlantic Project, and the US hegemony was created. The future of the Atlantic alliance is a key issue for the future of the world system

Wallerstein (1988) has recently speculated over the possible implications of European unity for the world system. He takes his departure in what he calls 'the normal entropy of monopolistic advantage within capitalism' or, in other words, the relative decline of the US. The most important European contribution to a reorganization of the world system would be continental unity. The main opening for this unity would be a global realignment, with a US–Japan–China alliance stimulating closer cooperation between Western and Eastern Europe and the Soviet Union. Thus Europe's scope for action depends both on global developments in the world system at large and the internal process of integration within Europe itself.

References

Addo H et al. (eds) 1985 *Development as social transformation*. Hodder and Stoughton, London
Adelman I, Taft Morris C 1973 *Economic growth and social equity in developing countries*. Stanford University Press, Stanford, California
Aganbegyan A 1988 *The challenge: economics of perestroika*. Hutchinson, London
Aguilar L E (ed) 1968 *Marxism in Latin American*. Alfred A Knopf, New York
Ake C 1979 *Social science in Nigeria*. Working paper, CODESRIA
Alatas S H 1972 The captive mind in development studies, *International Social Science Journal* **XXIV**, No. 4
Almond G A 1970 *Political development: essays in heuristic theory*. Little Brown, Boston
Almond G A, Coleman J (eds) 1960 *The politics of the developing areas*. Princeton University Press, Princeton
Almond G A, Powell C B 1965 *Comparative politics: a developmental approach*. Little Brown, Boston
Almond G A, Verba A 1963 *The civic culture*. Princeton University Press, Princeton
Amin S 1974 *Accumulation on a world scale*. Harvester Press, Sussex
Amin S 1976 *Unequal development*. Harvester Press, Sussex
Amin S 1977a *Imperialism and unequal development*. Harvester Press, Sussex
Amin S 1977b Universality and cultural spheres. *Monthly Review* Feb. 1977
Amin S 1980 Class and nation, historically and in the current crisis. MR Press, New York
Amin S 1985 A propos the 'Green movements'. In Addo 1985
Amin S, Arrighi G, Frank, A G, Wallerstein I 1982 *Dynamics of global crisis*. Monthly Review Press
Anderson P 1975 *Lineages of the absolutist state*. NLB, London
Anell L 1981 *Recession. The Western economies and the changing world order*. Frances Pinter, London
Angelopoulos A 1983 *Global plan for employment. A new Marshall plan*. Praeger, New York
Apter D 1965 *The politics of modernization*. University of Chicago Press, Chicago
Apter D E 1987 *Rethinking development. Modernization, dependency and postmodern politics*. Sage, Newbury Park
Arbatov G 1983 *Cold war or détente? The Soviet viewpoint*. Zed Books, London

Arndt H W 1981 Economic development: a semantic history. *Economic Development and Cultural Change* **29**(3) (April)

Arrighi G 1983 *The geometry of imperialism. The limits of Hobson's paradigm* Verso/NLB, London

Asad T (ed) 1973 *Anthropology and the colonial encounter*. Ithaca Press, London

Atal Y (ed) 1974 *Social sciences in Asia*. Abhinav Publications, New Delhi

Atal Y 1981 The call for indigenization. *International Social Science Journal* **32**(1): 189–97

Atal Y, Pieris R 1976 *Asian rethinking on development*. Abhinav Publications, New Delhi

Atta-Mills L 1979 *The role of social scientists in development: the rise, fall and rebirth of social science in Africa*. CODESRIA Working Paper No. 10, Dakar

Bablewski Z 1987 Evolution of the development theory of socialist countries (paper for EADI General conference, Amsterdam). Also in Bozky 1988

Bach R L 1980 On the holism of a world-systems perspective in Hopkins T K, Wallerstein I (eds), *Processes of the world system*. Sage Publications, Beverly Hills

Bahro R 1984 *From red to green*. Verso/NLB, London

Bahro R 1986 *Building the green movement*. Heretic Book, GMP Publishers, London

Balandier G 1951 La Situation coloniale: approche théorique. *Cahiers Internationaux de Sociologie* **XI**: 44–79

Bambirra V 1974 *El capitalismo dependiente Lationamericano*. Siglo XXII, México

Banerjee D (ed) 1985 *Marxian theory and the third world*. Sage, New Delhi

Baran P 1957 *The political economy of growth*. Monthly Review Press, New York and London

Barnett H J, Morse C 1963 *Scarcity and growth. The economics of natural resource availability*. John Hopkins, Baltimore

Barraclough G 1980 Worlds apart: untimely thoughts on development and development strategies. Discussion paper 152, Institute of Development Studies, Sussex

Bates R 1981 *Markets and states in Africa*. University of California Press, Berkeley

Bauer P T, Yamey B S 1957 *The economics of underdeveloped countries*. Cambridge University Press, Cambridge

Bauer P T 1954 *West African trade*. Cambridge University Press, Cambridge

Bauer P T 1971 *Dissent on development*. Weidenfeld and Nicolson, London

Bauer P T 1981 *Equality, the third world and economic delusion*. Harvard University Press, Cambridge Mass

Beaud M 1987 The crisis of development in the light of economic system analysis. *Review* **X**(3)

Beckerman W 1974 *In defence of economic growth*. Jonathan Cape, London

Beckford L G 1972 *Persistent poverty. Underdevelopment in plantation economies of the third world*. Oxford University Press, New York

Bell D 1973 *The coming of post-industrial society: a venture in social forecasting*. Basic Books, New York

Belshaw C S 1965 *Traditional exchange and modern markets*. Prentice Hall, Englewood Cliffs N.J.

Berger P L 1977 *Pyramids of sacrifice. Political ethics and social change*. Penguin, Harmondsworth

Berlin I 1979 *Russian thinkers*. Penguin, Harmondsworth

Bernstein H 1977 Underdevelopment and the law of value: a critique of Kay'. *Radical African Political Economy* 8

Bernstein H 1979 Sociology of underdevelopment versus sociology of development in Lehman D (ed) 1979

Bernstein R J (ed) 1985 *Habermas and modernity*. Polity Press, Cambridge

Bidelux R 1985 *Communism and development*. Methuen, London

Bienefeld M, Godfrey M (eds) 1982 *The struggle for development: National strategies in an international context*. John Wiley, London

Bienefeld M 1982 Tanzania: model or anti-model? In Bienefeld and Godfrey 1982

Bienefeld M 1987 The significance of the East Asian NICs for the development policy debate. Paper presented to the *Third Annual conference of the Canadian Association for the Study of International Development*. Hamilton, June 1987

Binder L et al. 1971 *Crisis and sequences in political development*. Princeton University Press, Princeton

Blaikie P 1985 *The political economy of soil erosion in developing countries*. Longman, London

Blomström M, Hettne B 1984 *Development theory in transition. The dependency debate & beyond*. Third world responses. ZED Books

Bonfil G 1982 *America Latina y Etnocidio*. FLACSO, San José

Bookchin M 1980 *Toward an ecological society*. Black Rose Books, Montreal

Bookchin M 1987 Thinking ecologically. A dialectical approach. *Our generation* **18**(2)

Booth D 1985 Marxism and development sociology: interpreting the impasse. *World Development* **13**(7): 761–87

Bossert T J 1987 The promise of theory. In Klarin P F, Bossert T J (eds) 1987 *Promise of development. Theories of change in Latin America*. Westview Press, Boulder, Colorado

Boulding K 1966 Economics and ecology. In Fraser-Darling F, Milton J D (eds) *Future environments of North America*. Natural History Press

Bozyk P 1988 *Global challenges and East European responses*. Polish Scientific Publishers, Warsaw

Brandt W 1986 *World armament and world hunger*. Victor Gollancz, London

Brenner R 1977 The Origins of capitalist development: a critique of neo-Smithian Marxism. *New Left Review* 104

Brookfield H 1975 *Interdependent development*. Methuen, London
Brown L R 1981 *Building a sustainable society*. Norton W W, New York and London
Brus W 1979 The East European reforms: what happened to them?. *Soviet Studies* **XXXI**(2) April 1979: 257–67
Brus W 1987 Experience of the socialist countries. In Emmerij 1987
Bull H, Watson A (eds) 1984 *The expansion of international society*. Oxford University Press
Burling R 1962 Maximization theories and the study of economic Anthropology. *American Anthropologist* **64**
Bury J B 1955 (1932) *The idea of progress. An inquiry into its origin and growth*. Dover Publications, New York
Buzan B 1984 Economic structure and international security. In *International Organization* **38**(4)
Cardoso F H 1967 The industrial elite. In Lipset S M, A Solari (eds). *Elites in Latin America*. Oxford University Press, New York
Cardoso F H 1976 The consumption of dependency theory in the U.S., Paper submitted to the *Third Scandinavian Research Conference on Latin America, Bergen, June 1976*
Cardoso F H 1977 The originality of a copy: CEPAL and the idea of development. CEPAL *Review* (second half of 1977): 7–40
Cardoso F H, Faletto E 1969 (1979) *Dependencia y Desarrollo en América Latina*, Mexico: Siglo XXI (English translation: Marjory Mattingly Urquidi 1979, *Dependency and Development in Latin America*. University of California Press, Berkley
Carr E H, Davies R W 1974 *A history of Soviet Russia. Foundations of a planned economy*, vol. 1. Pelican Books
Chambers R 1985 Putting 'Last' thinking first: a professional revolution. In Gauhar, Raana (ed) *Third World Affairs 1985*. Third World Foundation, London
Chandra S 1975 *Dependence and disillusionment. Emergence of national consciousness in later 19th century India*. Manas Publications, New Delhi
Chenery H 1975 Restructuring the world economy. *Foreign Affairs* **54**(2)
Chenery H et al. 1974 *Redistribution with growth*. Oxford University Press, London
Chilcote R H (ed) 1982 *Dependency and Marxism. Toward a resolution of the debate*. Westview Press, Boulder, Colorado
Chilcote R H 1984 *Theories of development and underdevelopment*. Westview Press, Boulder and London
Clammer J (ed) 1978 *The new economic anthropology*. Macmillan Press, London
Clapman C 1985 *Third world politics. An introduction*. Croom Helm, London
Clarkson S 1978 *The Soviet theory of development: India and the third world in Marxixt-Leninist scholarship*. University of Toronto Press
Cline W R 1982 Can the East Asian model of development be generalized?. *World Development* **10**(2): 81–90
Cockroft J D, Frank A G, Johnson D L (eds) 1972 *Dependence and underdevelopment. Latin America's political economy*. New York

Cockroft J D, Gandy R 1981 The Mexican volcano. *Monthly Review* **33**(1)
Collier D C 1979 *The new authoritarianism in Latin America*. Princeton University Press
Collins J, Lappé F Moore 1980 Food self-reliance. In Galtung, O'Brien and Preiswerk 1980
Corbridge S 1986 *Capitalist world development. A critique of radical development geography*. Macmillan, London
Cruise O'Brien D 1979 Modernization, order and the erosion of a democratic ideal. In Lehman 1979
Cueva A 1974 (1976) Problemas y perspectivas de la teoria de la dependencia, *Historia y Sociedad*, Núm. 3, México
Cumper G E 1974 Dependence, development and the Sociology of economic thought. *Social and Economic Studies* **23**(3)
Dahlström E 1988 Ideological orientation and political control of Vietnamese social science. *Acta Sociologica* 31:2: 105–17
Dalton G (ed) 1968 *Primitive archaic and modern economics. Essays of Karl Polanyi*. Doubleday, New York
Datta A 1987 Understanding East Asian economic development. *Economic and Political Weekly* **XXII**(14) 4 April 1987
Davies H E 1972 *Latin American thought: a historical introduction*. The Free Press, New York
de Kadt E, Williams G (eds) 1974 *Sociology and development*. Tavislock, London
de Rougemont D, 1980 The responsibilities of Europe *vis-à-vis* the NIEO: a minority view. In Laszlo and Kurtzman 1980
Demas W G 1965 *The economics of development in small countries with special reference to the Caribbean*. McGill University Press, Montreal.
Deutsch K W 1977 *Ecosocial systems and ecopolitics*. UNESCO, Paris
Dia M 1960 *Reflexions sur l'economie de l'Afrique Noir*. Présence Africaine, Paris
Dos Santos T 1970 The structure of dependency. *American Economic Review* **60**(21) May 1970
Dos Santos T 1977 Dependence relations and political development in Latin America: some considerations. *Ibero-Americana* **VII:I**
Drucker P F 1986 The changed world economy. *Foreign Affairs* **64**(4)
Ekins P (ed) 1986 *The living economy: a new economics in the making*. Routledge, London
Ekins P 1988 Green ideas on economics and security and their political implications. In Friberg 1988
Ellis H S, Wallich H C (eds) 1961 *Economic development for Latin America*. Macmillan, London
Emmerij L (ed) 1987 *Development policies and the crisis of the 1980s*. OECD, Paris
Emmerij L 1988 Peace and poverty: Europe's responsibility. In Hettne 1988a
Erlich A 1960 (2nd edn 1967) *The Soviet industrialization debate 1924–1928*. Harvard University Press, Cambridge Mass
Etzioni A 1968 *The active society: a theory of societal and political processes*. Collier–Macmillan, New York

Faber M, Seers D (eds) 1972 *The crisis in planning*. Chatto & Windus, London
Fagen R, Deere, C D, Coraggio, J L (eds) 1986 *Transition and development*. Monthly Review Press, New York
Falk R 1980 Normative initiatives and demilitarization: a third system approach. *Alternatives* **6**(2)
Fals Borda O 1970 *Ciencia Propia y Colonialismo Intelectual*. Nuestra Tiempo, Bogota
Fehér F, Heller A 1987 *Eastern left, Western left. Totalitarianism, freedom and democracy*. Polity Press, Cambridge
Fehér F, Heller A, Márkus G 1983 *Dictatorship over needs*. Basil Blackwell, Oxford
Ferrer A 1985 Argentina's foreign debt crisis. In: Gauhar, Raana (ed) *Third World Affairs 1985*. Third World Foundation, London
Firth R 1939 *Primitive Polynesian economy*. Routledge & Kegan Paul, London
Forbes D K 1984 *The geography of underdevelopment*. Croom Helm, London and Sydney
Fortin C 1984 The failure of repressive monetarism: Chile 1973–1983. *Third World Quarterly* **6**(1)
Foster-Carter A 1974 Neo-Marxist approaches to development and underdevelopment. In de Kadt and Williams (eds) 1974
Foster-Carter A 1976a From Rostow to Gunder Frank: conflicting paradigms in the analysis of underdevelopment. *World Development* 4; 3, 1976
Foster-Carter A 1976b Marxism and the 'Fact of Conquest'. *African Review* **6**(1)
Foster-Carter A 1978 The modes of production controversy. *New Left Review* (107) 56 ff
Foster-Carter A 1985 *The sociology of development*. Causeway Press, Ormskirk, Lancashire
Frank A G 1966 The development of underdevelopment. *Monthly Review* Sept: 17–30
Frank A G 1969 *Latin America: underdevelopment or revolution*. Monthly Review Press, New York
Frank A G 1972 *Lumpenbourgeoisie–Lumpendevelopment*. Monthly Review Press, New York
Frank A G 1977 Dependence is dead, long live dependence and the class struggle: an answer to critics. *World Development* 5(4) April 1977
Frank A G 1981 *Reflections on the world economic crisis*. MR Press, New York
Frank A G 1983a *The European challenge. From Atlantic Alliance to pan-European entente for peace and jobs*. Spokesman, Nottingham
Frank A G 1983b Global crisis and transformation. *Development and Change* **14**(3) July 1983. Also in Frank 1984
Frank A G 1984 *Critique and anti-critique. Essays on dependence and reformism*. Macmillan, London
Freeman C (ed) 1984 *Long waves in the world economy*. Frances Pinter, London and Dover
Friberg M (ed) 1988 *New social movements in Europe*. United Nations

University European Perspectives Project, Gothenburgh: Padrigu Papers
Friberg M, Hettne B 1985 The greening of the world: towards a non-deterministic model of global processes. In Addo et al. (eds) 1985
Friberg M, Hettne B, Tamm G 1979 *Societal change and development thinking: an inventory of issues*. United Nations University GPID project, Tokyo
Friedland W, Rosberg C Jun. (eds) 1964 *African Socialism*. University Press, Stanford
Friedman J, Weaver C 1979 *Territory and function: the evolution of regional planning*. Edward Arnold, London
Friedman M 1962 *Capitalism and freedom*. University of Chicago Press, Chicago
Fröbel F, Heinrichs J, Kreye O 1980 *The new international division of labour*. Cambridge University Press, Cambridge
Furtodo C 1963 *The economic growth of Brazil*. University of California Press, Berkeley
Galanter M 1984 Competing equalities. Law and the backward classes in India. Oxford University Press, Delhi
Galtung J, O'Brien P, Preiswerk R (eds) 1980 *Self-reliance, a strategy for development*. Bogle-L'Ouuverture, London
Galtung J (n.d.) *Two ways of being western: some similarities between Liberalism and Marxism* (mimeo). Chair in Conflict and Peace Research, Paper No. 97 Oslo,
Galtung J 1976 *The politics of self-reliance*, (mimeo). Chair in Conflict and Peace Research, Paper No. 44 Oslo,
Galtung J 1977 *Human needs as the focus of social sciences* (mimeo). Chair in Conflict and Peace Research, Paper No. 51 Oslo,
Galtung J 1985 Development theory: notes for an alternative approach. Institute for Environment and Society, Berlin
Galtung J, O'Brien P, Preiswerk R (eds) 1980 *Self-reliance: a strategy for development*. Bogle-L'Overture Publications, London
Gamble A 1985 *Britain in decline*. Macmillan, London
Gareau F H 1987 Expansion and increasing diversification of the universe of social science. *International Social Science Journal* (114)
Garton Ash T 1985 *The Polish revolution*. Coronet Books, Hodder & Stoughton, London
Gellner E 1964 *Thought and change*. Weidenfeld and Nicolson, London
George S 1985 From the world food conference to 1984: a decade of failure. *Third World Affairs 1985*. Third World Foundation, London
George S 1988 *A fate worse than debt*. Penguin, London
Gerschenkron A 1962 *Economic backwardness in historical perspective*. Harvard University Press, Cambridge, Mass.
Gershuny J I 1979 The informal economy. Its role in postindustrial society. *Futures* Feb. 1979
Girvan N 1973 The development of dependency economics in the Caribbean and Latin America: review and comparison. *Social and Economic Studies* 22(1)
Glaeser B 1984 *Ecodevelopment: concepts, projects, strategies*. Pergamon Press, Oxford

Godelier M 1987 Introduction: the analysis of transition processes. In *International Social Science Journal* (114) Nov. 1987
Gonzales Casanova P 1965 *La Democracia en México.* Siglo XXI Mexico City,
Goulborne H (ed) 1979 *Politics and the state in the third world.* Macmillan, London
Green R H 1978 Basic human needs: concept or slogan, synthesis or smoke screen. *IDS Bulletin* **9**(4)
Griffin K 1988 *Alternative strategies for economic development* (manuscript), OECD
Griffith Jones S 1988 External debt and development in Latin America: the last decade of the eighties and perspectives for the nineties, (Paper for the *8th Nordic Research Conference on Latin America, Stockholm, July 1988*)
Gunasinghe N 1984 The open economy and its impact on ethnic relations in Sri Lanka. In *Sri Lanka. The Ethnic Conflict*, Committee for Rational Development. Navrang Publishers, New Delhi
Gurley J 1976 *China's economy and the Maoist strategy.* Monthly Review Press, New York and London
Gurley J 1978 Economic development: a Marxist view. In Jameson and Wilber 1978
Gurrieri A The validity of the state-as-planner in the current crisis. CEPAL *Review* (31) (April 1987)
Habermas J 1976 *Legitimation crisis.* Heinemann, London
Hadjimichalis C 1987 *Uneven development and regionalism. State, territory and class in southern Europe.* Croom Helm, London
Hager W 1985 'Fortress Europe': a model?. In Musto and Pinkele 1985
Haggard S 1986 The newly industrializing countries in the international system. *World Politics* **XXXVIII**(2) Jan 1986
Hall S, Jacques M (eds) 1983 *The politics of Thatcherism.* Lawrence and Wishart, London
Hamilton N 1982 *The limits of state autonomy. Post-revolutionary Mexico.* Princeton University Press
Harris N 1987 *The end of the third world. Newly industrializing countries and the decline of an ideology.* Penguin Books
Harrison R J 1974 *Europe in question: theories of regional international integration.* George Allen & Unwin, London
Hartmann J 1988 President Nyerere and the state. In M Hodd (ed) *Tanzania after Nyerere.* Francis Pinter, London
Hettne B 1976 The vitality of Gandhian tradition. *Journal of Peace Research* 3, **XIII**, 227–45
Hettne B 1978 *The political economy of indirect rule. Mysore 1881–1947.* Curzon Press, London
Hettne B 1983 Self-reliance and destabilization in the Caribbean and central America: the cases of Jamaica and Nicaragua'. *Current Research on Peace and Violence*, **VI** No 2–3
Hettne B 1984 *Approaches to the study of peace and development.* EADI Working Papers, Tilburg
Hettne B 1985 The Ghanaian experiments with military rule. In Wallensteen, Galtung and Portales 1985

Hettne B 1986 An inventory of European social projects for peace and development. In Làszlo E, *Europe in the Contemporary World*. Gordon and Breach, New York

Hettne B (ed) 1988a *Europe, dimensions of peace*. ZED Books, London

Hettne B (ed) 1988b *Development options in Europe*, UNU: European Perspectives Project. Padrigu Papers, Gothenburg

Hettne B, Hveem H (eds) 1988 *Regionalism and interregional relations*. Padrigu Papers, Gothenburg

Higgins B 1980 The disenthronement of basic needs? Twenty questions. *Regional Development Dialogue* **1**(1)

Higgot R A 1983 *Political development theory*. Croom Helm, London

Hilton R (ed) 1978 *The transition from feudalism to capitalism*. Verso, London

Hirsch F 1976 *Social limits to growth*. Harvard University Press, Cambridge, Massachusetts

Hirschman A O (ed) 1961 *Latin American issues – essays and comments*. Twentieth Century Fund, New York

Hirschman A O 1978 *The passions and the interests. Political arguments for capitalism before its triumph*. Princeton University Press, Princeton, N.J.

Hirschman A O 1981 *Essays in trespassing: economics to politics and beyond*. Cambridge University Press

Hirschman A O 1985 *The strategy of economic development*. Yale University Press

Hobsbawn E J 1964 Introduction to Karl Marx. *Precapitalist Economic Formations*. Lawrence and Wishart, London

Hodgkin T 1972 Some African and third world theories of imperialism. In Owen and Sutcliffe 1972

Holland S 1983 *Out of crisis. A project for European recovery* (Forum for International Political and Social Economy). Spokesman, Nottingham

Hollist W L, Rosenau J N 1982 *World system structure. Continuity and change*. SAGE, Beverly Hills and London

Holloway D 1981 *War, militarism and the Soviet state*. Institute for World Order, New York

Hoogvelt A M 1982 *The third world in global development*. Macmillan, London

Hopkins T K, Wallerstein I 1987 Capitalism and the incorporation of new zones into the world economy. *Review* **X**(5/6): 763–79

Hoselitz B F (ed) 1960 *The sociological aspects of economic growth*. Free Press, New York

Hübner W 1988 The Polish way of reform. In Hettne 1988b

Hueting R 1980 *New scarcity and economic growth*. North Holland Publishing Company, Amsterdam

Huntington S P 1968 *Political order in changing societies*. Yale University Press, New Haven, Conn.

Huntington S P 1971 The change to change. Modernization, development and politics. *Comparitive Politics* **3** No 3

Hveem H 1976 The politics of the new international economic order. *Bulletin of Peace Proposals* **7** No 1

Hveem H 1980 Scandinavia, the like-minded countries and the NIEO. In Laszlo and Kurtzman 1980
Hydén G 1983 *No shortcuts to progress.* University of California Press, Berkeley
Hydén G 1984 Ethnicity and state coherence in Africa. *Ethnic Studies Report* **II**(1) Jan. 1984
ILO 1976 *Declaration of principles and programme of action for a basic needs strategy of development* (adopted by the participants in the 1976 conference on World Employment) International Labour Office, Geneva
Ionescu G, Gellner E 1969 *Populism. Its meanings and national characteristics.* Weidenfeld & Nicolson, London
Jameson K, Wilber C (eds) 1978 *Directions in economic development.* University of Notre Dame Press, Notre Dame
Jasny N 1972 *Soviet economists of the twenties.* Harvard University Press, Cambridge, Mass.
Johnston R J, Taylor P J (eds) 1986 *A world in crisis?.* Basil Blackwell, Oxford and New York
Jorrin M, Martz J D 1970 *Latin American political thought and ideology.* University of North Caroline Press, Chapel Hill
Judy R W 1971 The economists. In H G Skilling, F Griffiths *Interest Groups in Soviet Politics.* Princeton University Press
Kahn H et al. 1976 *The next 2000 years: a scenario for America and the world.* William Morrow, New York
Kaplinsky R 1984 *Third world industrialization in the 1980s. Open economies in a closing world.* Frank Cass, London
Kay G 1975 *Development and underdevelopment: a Marxist analysis.* Macmillan, London
Keynes J M 1933 National self-sufficiency. *Yale Review* **XXII**(4): 755–69
Khor Kok Peng (1983) *Recession and the Malaysian economy.* Institut Masyarakat, Penang
Khoros V 1984 *Populism: its past, present and future.* Progress Publishers, Moscow
Killick T 1978 *Development economics in action. A study of economic policies in Ghana.* Heineman, London
King D S 1987 *The new right. Politics, markets and citizenship.* MacMillan, London
Kitching G 1982 *Development and underdevelopment in historical perspective: populism, nationalism and industrialization.* Methuen, London and New York
Klinghoffer A J 1969 *Soviet perspectives on African socialism.* Rutherford/Fairleigh Dickinson University Press, Cranbury, N.J.
Kofi T A 1974 Development and stagnation in Ghana: an 'Abibirim' approach. *Universitas* **3** No 2.
Kofi T A 1975 Towards the 'Abibirim' strategy of development: a formal articulation. *Universitas* 5 No 1
Korany B 1986 Hierarchy within the South: in search of theory. In: Gauhar, Raana (ed) *Third World Affairs.* London
Kornai J 1980 *Economics of shortage.* North Holland, Amsterdam
Kothari R 1970 *Politics in India.* Little Brown, Boston

Kothari R 1984 Communications for alternative development: towards a paradigm. *Development dialogue* 1984: 1–2

Krasner S D 1985 *Structural conflict: the third world against global liberalism*. University of California Press, Berkeley, Los Angeles, London

Kridl Valkenier E 1980 Development issues in recent soviet scholarship. *World Politics* **XXXII**(4)

Kridl Valkenier E 1983 *The Soviet Union and the third world: an economic bind*. Praeger, New York

Krueger A O 1986 Changing perspectives on development economics and World bank research. *Development Policy Review* **4**(3)

Kumar K 1978 *Prophecy and progress. The sociology of industrial and postindustrial society*. Penguin, Harmondsworth

Laclau E 1971 Feudalism and capitalism in Latin America. *New Left Review* (67)

Lal D 1983 *The poverty of development economics*. The Institute of Economic Affairs, London

Lall S 1975 Is dependence a useful concept in analysing underdevelopment. *World Development* (11) 1975

Lamb G 1981 Rapid capitalist development models: a new politics of dependençe. In D Seers *Dependence theory: A critical assessment*. Frances Pinter, London

Laszlo E, Kurtzman J (eds) 1980 *Western Europe and the new international economic order*. Pergamon Press, New York

Lederer K (ed) 1980 *Human needs. A contribution to the current debate*. Oelgeschlager, Gunn & Hain/Verlag Anton

Lehmann D (ed) 1979 *Development theory. Four critical studies*. Frank Cass, London

Lehmann D 1986 Dependencia: an ideològical history. *Discussion Paper 219*, Institute of Development Studies, Sussex

Leibenstein H 1957 *Economic backwardness and economic growth*. Wiley, New York

Lenin V I 1967 *The development of capitalism in Russia*. Progress Publishers, Moscow

Lerner D 1962 *The passing of the traditional society*. Free Press, Glencoe

Levitas R 1986 *The ideology of the new right* Polity Press, Cambridge

Levitt K 1970 *Silent surrender. The multinational corporation in Canada* Gage, Toranto and Ontario

Lewin M 1975 *Political undercurrents in Soviet economic debates*. Pluto Press, London

Lewis W A 1954 Economic development with unlimited supplies of labour. *The Manchester school of Economic and Social Studies* **XXII**(2) May 1954

Lewis W A 1950 Industrialization of the British West Indies. *Caribbean Economic Review* **2**(1)

Lewis W A 1955 *The theory of economic growth*. George Allen and Unwin, London

Leys C 1975 *Underdevelopment in Kenya. The political economy of neo-colonialism*. Heinemann, London

Leys C 1977 Underdevelopment and dependency: critical notes. *Journal of Contemporary Asia* **7**(1)

Liepitz A 1987 *Mirages and miracles. The crises of global Fordism* Verso, London

Lindberg L N (ed) 1976 *Politics and the future of industrial society.* David McKay, New York

Lipton M 1977 *Why poor people stay poor. A study of urban bias in world development* Temple Smith, London

List F 1841 *Das Nationale System der Politischen Ökonomie.* [In English: *The National system of Political Economy.* Longman 1985, London]

Little I M 1982 *Economic development. Theory, policy and international relations.* Basic Books, New York

Livingstone I (ed) 1971 *Economic policy for development: selected readings.* Penguin, Harmondsworth

Loney M 1986 *The politics of greed: the new right and the welfare state.* Pluto Press, London

Love J L 1980 Raúl Prebisch and the origins of the doctrine of unequal exchange. *Latin American Research Review* **15**: 1

Lukey F W 1978 *Independence and into the 21st century – an appraisal of the future in Ghana.* UST, Kumasi

McClelland D 1962 *The achieving society.* Van Nostrand, Princeton

MacFarquhar R (ed) 1974 *The hundred flowers campaign and the Chinese intellectuals.* Octagon Books, New York

McGowan P, Johnson H 1984 Sixty coups in thirty years – further evidence regarding African military coups d'etat'. *The Journal of Modern African Studies* **24**(3)

Malinowski B 1961 *Argonauts of the Western Pacific* E P Dutton, New York

Mandaza I 1988 The relationship of third world intellectuals and progressive western scholars: an African Critique. *SAPEM* February 1988

Mandel E 1964 The heyday of neo-capitalism and its aftermath. *Socialist Register*

Mandel E 1984 Explaining long waves of capitalist development. In Freeman 1984

Mandel E 1986 In defence of socialist planning. *New Left Review* (159) Sept./Oct 1986

Mandelbaum K 1945 *The industrialization of backward areas* Basil Blackwell, Oxford

Mansbach R W, Ferguson Y H, Lampert, D E 1976 *The web of world politics: non-state actors in the global system.* Prentice Hall, Englewood Cliffs, N.J.

Marien M 1976 *Societal directions and alternatives: a critical guide to the literature.* New York

Marien M 1977 The two visions of postindustrial society. *Futures* Oct. 1977

Marini R M 1973 *Dialectica de la Dependencia.* Era, México

Mathias P 1969 *The first industrial nation: an economic history of Britain.* Methuen, London

Mattick P 1981 *Economic crisis and crisis theory.* Merlin Press, London

Max-Neef M 1986 Human scale economics: the challenges ahead. In Ekins 1986

Meier G M, Seers D (eds) 1984 *Pioneers in development*. Oxford University Press, New York
Meier G M 1987 *Pioneers in development (second series)*. Oxford University Press, New York
Meisner M 1979 Leninism and Maoism: some populist perspectives on Marxism-Leninism in China. *The China Quarterly* (Jan./March)
Melotti U 1977 *Marx and the third world*. Macmillan, London
Mendel A P 1961 *Dilemmas of progress in Tsarist Russia. Legal populism and legal Marxism*. Harvard University Press, Cambridge, Mass.
Mendlovitz S H (ed) 1975 *On the creation of a just world order. Preferred worlds for the 1990s*. The Free Press, New York
Mills C Wright 1959 *The sociological imagination*. Oxford University Press, New York
Milward, Alan S 1988 The origins of the Treaty of Rome. In Hettne 1988b
Minocha M 1970 Drain theory and its relevance to present day trade relations between developed and under-developed countries and rural–urban sectors. In *The drain theory, papers read at the Indian economic conference*. Popular Prakash, Bombay
Mishan E J 1967 *The costs of economic growth*. London
Mishra R 1984 *The welfare state in crisis*. Harvester Press
Mkandawire T 1988 Problems of social sciences in Africa. Paper for *SID World Conference, New Delhi, March 1988*
Moore B 1966 *The social origins of democracy and dictatorship*. Beacon Press, Boston
Moore W E 1977 Modernization as rationalization: process and restraints. In Nash 1977
Musto S (ed) 1985 *Endogenous development. A myth or path? Problems of economic self-reliance in the European periphery*. European Association for Development Research and Training Institutes, Tilburg
Musto S, Pinkele C F 1985 *Europe at the crossroads – agendas of the crisis*. Praeger, New York
Myrdal G 1956 *An international economy*. Harper, New York
Myrdal G 1957 *Economic theory and underdeveloped regions*. Ducksworth, London
Myrdal G 1968 *Asian drama* Pantheon, New York
Naoroji D 1962 *Poverty and un-British rule in India*. Publications Divisions, Ministry of Information and Broadcasting, Government of India, Delhi
Nash M (ed) 1977 *Essays on economic development and cultural change in honor of Bert F Hoselitz*. University of Chicago Press, Chicago
Nash M 1984 *Unfinished agenda: the dynamics of modernization in developing nations*. Colorado: Westview Press, Boulder
Nayar B R 1972 *The modernization imperative of Indian planning*. Vikas, Delhi
Nerfin M (ed) 1977 *Another development: approaches and strategies*. The Dag Hammarskjöld Foundation, Uppsala
Nettleford R 1978 *Caribbean cultural identity. The case of Jamaica*. The Institute of Jamaica, Kingston
Nisbet R 1969 *Social change and history*. Oxford University Press, London

Nisbet R 1980 *History of the idea of progress*. Heinemann Education Books, London
Nixson F, n.d. Beyond economic development. Manchester Discussion Papers in Development Studies, University of Manchester
Nove A 1964 *Was Stalin really necessary?* George Allen and Unwin, London
Nove A 1983 *The economics of feasible socialism*. George Allen & Unwin, London
Nugent J B, Yotopoulos P B 1979 What has orthodox development economics learned from recent experience. *World Development* 7(6)
Nurkse R 1953 *Problems of capital formation in underdeveloped countries*, Oxford: Basil Blackwell.
Nyerere J K 1968 *Ujamaa: Essays on socialism*. Dar Es Salaam
O'Connor J 1987 *The meaning of crisis*. Basil Blackwell, Oxford
O'Donnel G 1973 *Modernization and bureaucratic authoritarianism: studies in South American politics*. Institute of International Studies (University of California), Berkeley
Oommen T K, Mukherji R N (eds) 1986 *Indian sociology: reflections and introspections*. Popular Prakashan, Bombay
Owen R, B Sutcliffe (eds) 1972 *Studies in the theory of imperialism*. Longman, London
Oxaal I, Barnett, I D, Booth, D 1975 *Beyond the sociology of development*. Routledge & Kegan Paul, London
Pachter H M 1974 the idea of progress in Marxism. *Social Research* 41: 1
Palma G 1978 Dependency: a formal theory of underdevelopment or a methodology for the analysis of concrete situations of underdevelopment? *World Development* **6**(7/8)
Parsons H L 1977 *Introduction to Marx and Engels on ecology*. Greenwood Press, Westport and London
Parsons T 1951 *The social system* Free Press, Glencoe, Ill.
Payer C 1985 The IMF in the 1980s: what have we learned about it? In Gauhar, Raana (ed), *Third World Affairs 1985*. Third World Foundation, London
Paz O 1972 *The other Mexico: critique of the pyramid*. Grove Press, New York
Perroux F 1971 Note on the concept of growth poles. In Livingstone 1971 (1955)
Perroux F 1950 Economic space: theory and applications. *Quarterly Journal of Economics* (64): 89–104
Pirages D 1978 *The new context for international relations: global ecopolitics*. Duxbury Press, North Scituate; Massachusetts
Plowiec U 1978 Problems of foreign trade management at the stage of transition to open economy. *Economic Papers* No. 7, 1978. Central School of Planning and Statistics, Warsaw
Polanyi K 1945 Universal capitalism or regional planning. *The London Quarterly of World Affairs* Jan. 1945
Polanyi K 1957 *The great transformation*. Beacon Press, Boston
Polanyi K 1977 *The livelihood of man* (ed. by H. Pearson). Academic Press

Polanyi K, Arensberg C M, Pearson H W (eds) 1957 *Trade and market in the early empires*. Free Press, Glencoe, Ill.

Pollock D H 1978 Some changes in United States' attitudes towards CEPAL over the past 30 years. *CEPAL Review* No. 9, 1980

Prebisch R 1950 *The economic development of Latin America and its principal problems*. United Nations, New York

Prebisch R 1980 Towards a theory of change. *CEPAL Review* No 9, 1980

Preobrazhensky E 1965 *The new economics*. Clarendon Press, Oxford

Preston P W 1982 *Theories of development*. Routledge & Kegan Paul, London

Preston P W 1985 *New trends in development theory*. Routledge & Kegan Paul, London

Pye L W 1966 *Aspects of political development*. Little Brown, Boston

Randall V, Theobald R 1985 *Political change and underdevelopment*. Macmillan, London

Ray D I 1986 *Ghana, politics, economics and society*. Frances Pinter, London

Redclift M 1987 *Sustainable development. Exploring the contradictions*. Methuen, London and New York

Ries C 1977 The 'New international economic order': the Sceptics' Views. In Sauvant and Hasenpflug 1977

Riggs F 1964 *Administration in developing countries: the theory of prismatic society*. Houghton Mifflin, Boston

Riggs F 1981 The rise and fall of political development. In S C Long (ed) *The handbook of political behaviour, Vol. 4*. Plenum Press, New York and London

Riskin C 1987 *China's political economy: the quest for development since 1949*. Oxford University Press

Robbins L 1932 *An essay on the nature and significance of economic science*. Macmillan, London

Rodriguez O 1980 *La teoria del subdesarrollo de la CEPAL*, México: Siglo XXI

Rosenstein-Rodan P N 1961 Notes on the theory of the big push. In Ellis and Wallich 1961

Rosenstein-Rodan P N 1943 Problems of industrialization of eastern and south-eastern Europe. *Economic Journal* **53** June–Sept.

Ross S R 1975 *Is the Mexican revolution dead*? Philadelphia

Rostow W W 1960 *The stages of economic growth*. Cambridge University Press

Rostow W W 1984 *The barbaric counter-revolution. Cause and cure*. Macmillan, London

Rothchild D, Curry R L 1978 *Scarcity, choice and public policy in middle Africa*. University of California Press, Berkeley

Rothstein R L 1976 The political economy of redistribution and self-reliance. *World Development* **4**(7)

Roxborough I 1979 *Theories of underdevelopment*. Macmillan Press, London

Roy M N 1971 *India in transition*. Bombay

Rudolph L, S Rudolph 1967 *The modernity of tradition. Political development in India*. University of Chicago Press, Chicago

Rudqvist A 1986 *Peasant struggle and action research in Colombia*. Department of Sociology, Uppsala
Sachs I 1974 Ecodevelopment. *Ceres* **17** No 4: 17–21
Sachs I 1980 *Stratégies de l'écodévelopment*. Les Éditions Ouvriéres, Paris
Sahlins M 1972 *Stone age economics*. Aldine-Atherton, Chicago
Sandbrook R 1982 *The politics of basic needs. Urban aspects of assaulting poverty in Africa*. Heinemann, London
Sauvant K P, Hasenpflug H (eds) 1977 *The new international economic order. Confrontation or cooperation between north and south?* Witton House Publications, London
Schultz W T 1980 The economics of being poor. *Journal of Political Economy* **88**(4) August 1980
Seers D 1963 The limitations of the special case. *Bulletin of the Oxford Institute of Economics and Statistics* **25**(2) May 1963
Seers D 1978 *The congruence of Marxism and other neoclassical doctrines*. Discussion Paper, Institute of Development Studies, Sussex, August 1978
Seers D 1979a Patterns of dependence. In Villamil 1979
Seers D 1979b The birth, life and death of development economics. *Development and Change* **10**(4) Oct 1979
Seers D et al., 1979 *Underdeveloped Europe. Studies in core–periphery relations*. Harvester Press, Sussex
Seers D 1981 Development options: the strength and weaknesses of dependency theories in explaining a government's room to manoeuvre. In Seers D (ed), 1981
Seers D (ed) 1981 *Dependency theory: a critical assessment*. Frances Pinter, London
Seers D 1983 *The political economy of nationalism*. Oxford University Press
Sen A K 1981 *Poverty and famines: an essay on entitlement and deprivation*. Clarendon Press, Oxford
Sen G 1984 *The military origins of industrialization and international trade rivalry*. Frances Pinter, London
Senghaas D 1985 *The European experience*. Berg Publishers, Leamington Spa/Dover, New Hampshire
Shearman P 1987 Gorbachev and the third world: an era of reform? *Third World Quarterly* **9**(4) Oct 1987
Shiels F L (ed) 1984 *Ethnic separatism and world politics*. University Press of America, Lanham
Singer H W 1978 *The strategy of international development. Essays in the economics of backwardness*, (ed. by Sir Alec Cairncross and Mohinder Puri). Macmillan, London
Singer H W 1986 Raúl Prebisch and his advocacy of import substitution. *Development & South–South Cooperation* **II**(3) Dec 1986
Singh N 1988 The Apean Way. In Hettne 1988b
Skocpol T (ed) 1984 *Vision and method in historical sociology*. Cambridge University Press, Cambridge
Skocpol T 1979 *States and social revolutions*. Cambridge University Press, Cambridge

Smith A D 1973 *The concept of social change. A critique of the functionalist theory of social change*. Routledge & Kegan Paul, London

Smith M 1984 *Western Europe and the United States: the uncertain alliance*. George Allen & Unwin, London

Soja E 1983 Territorial idealism and the political economy of regional development. *City and Region* (6): 55–74

Solodovnikov V, V Bogoslovsky 1975 *Non-capitalist development. An historical outline*. Progress Publishers, Moscow

Srinivas M N 1968 Reflections on the study of one's own society. In *Social Change in Modern India*. University of California press, Berkeley, California

Stavenhagen R 1966 Siete tesis equivocados sobre America Latina. *Desarrollo Indoamericano* (4) 1966

Stavenhagen R 1986 Ethnodevelopment: a neglected dimension in Development thinking. In Anthorpe R, A Kráhl *Development studies: critique and renewal*. E J Brill, Leiden

Stepan A 1978 *The state and society: Peru in comparative perspective*. Princeton University Press

Street J 1987 The institutionalist theory of economic development. *Journal of Economic Issues* **XXI**(4) Dec 1987

Streeten P 1974 The role of social sciences in development studies. In *The Social Sciences and Development*. International Bank of Restruction and Development, Washington

Streeten P 1979 A basic-needs approach to economic development. In K P Jameson, C K Wilber (eds) *Directions in Economic Development*. University of Notre Dame Press, 1979, Notre Dame

Streeten P 1981 *Development perspectives*. Macmillan, London

Streeten P 1983 Development dichotomies. *Discussion Paper 187*, Institute of Development Studies, Sussex

Streeten P 1984 Interdependence: a north–south perspective. *Development and Peace* 5(1)

Stöhr W (ed) 1988 *Global challenge and local response. Learning experiences of local development initiatives*. United Nations University, European Perspectives Project, Tokyo

Sunkel O 1969 National development policy and external dependency in Latin America. *Journal of Development Studies* **1**(1)

Sunkel O 1973 Transnational capitalism and national disintegration in Latin America, *Social and Economic Studies* **22**(1)

Sunkel O 1986 The transnationalization of the centre–periphery system: in honour of Raúl Prebisch. *Development & South–South Cooperation* **II**(3) Dec 1986

Sunkel O, Fuenzalida E F 1979 Transnationalization and its national consequences. In Villamil 1979

Sunkel O, Paz P 1970 *El subdesarrollo latinoamericano y la teoria del desarrollo*. Siglo XXI, México

Sutton M 1985 Structuralism: the Latin American record and the new critique. In T Killick (ed) *The IMF and Stabilization: Developing Country Experiences*. Gower, in association with Overseas Development Institute, London

Taylor J (1974) Neo-Marxism and underdevelopment. *Journal of Contemporary Asia* (1) 1974

Taylor K 1982 *The political ideas of the utopian socialists*. Frank Cass, London

Taylor P J 1986 The world-system project. In R J Johnsson, P J Taylor *A world in crisis? Geographical perspectives*. Basil Blackwell, Oxford and New York

Temu P E 1975 Reflections on the role of social scientists in Africa. *International Social Science Journal* **XXVII**(1)

Thomas C 1974 *Dependence and transformation. The economics of the transition to socialism*. Monthly Review Press, New York

Thompson B R, Fogel K W 1976 *Higher education and social change* **1–2**. Praeger, New York

Thompson B R, Ronen D 1986 *Ethnicity, politics and development*. Lynn Rienner, Boulder, Colorado

Tilly C 1975 *The formation of national states in western Europe*. Princeton University Press, Princeton

Tilly C 1985 War and the power of warmakers in western Europe and elsewhere, 1600–1980. In Wallensteen, Galtung and Portales 1985

Timberlake L, J Tinker 1984 *Environment and conflict: links between ecological decay, environmental bankruptcy and political and military instability*. Earthscan, London.

Timberlake L 1985 *Africa in crisis*. Earthscan, London

Todaro M P 1977 *Economic development in the third world*. Longman, London

Towards one world? International responses to the Brandt Report 1981 Ed by the Friedrich-Ebert Foundation, Temple Smith, London

Toye J 1987a *Dilemmas of development. Reflections on the counter-revolution in development theory and policy*. Basil Blackwell, Oxford

Toye J 1987b Development theory and the experience of development. Issues for the future. In Emmerij 1987

Trimberger E K 1978 *Revolution from above*. Transaction Books, New Brunswick

Uberoi J P Singh 1968 Science and Swaraj. *Contributions to Indian Sociology* (new series) 1968: 2

Uchendu V C (ed) 1980 *Dependency and underdevelopment in West Africa*. Brill, Leiden

van Benthem van den Berg G 1972 Science and the development of society. Discussion paper (The Netherlands Universities Foundation for International Cooperation)

Vidyarthi L P (ed) 1970 *Gandhi and social sciences*. Bookhive, New Delhi

Villamil J J (ed) 1979 *Transnational capitalism and national development. New perspectives on dependence*. Harvester Press, IDS, Sussex

Villamil J J 1977 Only half a solution. *Mazingira* (3/4) 1977

Viner J 1953 *International trade and economic development*. Oxford University Press

von Laue T H 1963 *Sergei Witte and the industrialization of Russia*. New York

Walicki A 1969 *The controversy over capitalism. Studies in the social philosophy of the Russian populists*. Clarendon Press, Oxford

Wallensteen P J Galtung, Portales C (eds) 1985 *Global militarization*.

Westview Press, Boulder and London
Wallerstein I 1974 *The modern world system, capitalist agriculture and the origins of the European world economy in the sixteenth century.* Academic Press, New York and London
Wallerstein I 1976 The three stages of African involvement in the world-economy. In P C W Gutkind, I Wallerstein (eds), *The political economy of contemporary Africa.* Sage Publications, Beverly Hills/London
Wallerstein I 1979 *The capitalist world economy.* Cambridge University Press
Wallerstein I 1980 *The modern world system II. Mercantilism and the consolidation of the European world economy, 1600–1750.* Academic Press, New York and London
Wallerstein I 1988 European unity and its implications for the interstate system. In Hettne 1988a
Ward B, Dubos R 1972 *Only one earth.* Norton, New York
Warren B 1973 Imperialism and capitalist industrialization. *New Left Review* (81)
Warren B 1980 *Imperialism. Pioneer of Capitalism* New Left Books, London
Weeks J, Dore E 1979 International exchange and the causes of backwardness. *Latin American Perspectives*, Issue 21, **VI**(2) spring 1979
Weiner M (ed) 1966 *Modernization* Basic Books, New York
Weinstein M A 1976 *The polarity of Mexican thought: instrumentalism and finalism.* Pennsylvania State University Press, University Park and London
Weintraub A, Schwartz E, Aronson J R (eds) 1973 *The economic growth controversy.* International Arts & Sciences Press, New York
Werblan A 1988 Democratization in the political life: practical experiences. In Bozyk 1988
Werker S 1985 Beyond the dependency paradigm. *Journal of Contemporary Asia* **15**(1)
What now? Another development. The 1975 Dag Hammarskjöld Report. *Development Dialogue* (1/2) 1975
White G, et al. (eds) 1983 *Revolutionary socialist development in the third world.* Wheatsheaf Books, Brighton
Wilczynski W 1986 Polish economic reform and the economic theory of socialism. (*Œconomica Polona* **XIII**(1)
Wolf E R 1982 *Europe and the people without history.* University of California Press, Berkely
Wong, Siu-lun 1975 Social enquiries in the People's Republic of China. *Sociology* **9**(3) Sep 1975
Wong, Siu-lun 1979 *Sociology and socialism in contemporary China.* Routledge & Kegan Paul, London
Wood G (ed) 1985 *Labelling in development policy. Essays in honour of Bernard Schaffer.* Schaffer, London
Woodcock G 1975 *Anarchism.* Penguin Books
World Commission on Environment and Development 1987 *Our common future.* Oxford University Press
Worsley P 1984 *The three worlds. Culture and world development.* Weidenfeld and Nicolson, London

Index